The Self Publishing Toolbox

Tools, Tips, and Strategies for Becoming a Successful Author

Eric Frick

Published by Eric Frick 2020

Last Update January 2020

Copyright

Dedication

This book is dedicated to my Father Roy K. Frick. He was a great inspiration in my life and a role model for everyone who knew and worked with him. Although he passed away in 2008, his influence on his family, students and co-workers will leave a lasting legacy.

Table of Contents

1.0 Introduction

1.1 Book Introduction

Hello and welcome to The Self Publishing Toolbox. In this book, I will outline some of the tips and strategies that have worked for me to become a self-published author. I will share with you some of the tools that I've used that have made the process substantially easier than when I first started. I will also consolidate a lot of the information that I found in various places on the internet that has helped me become a much more effective author. I hope that after reading this book you will have a great start with your self-publishing business and that the tools that I give you a full save you a lot of time as well as some aggravation.

To become a successful author, it will require you to develop a great deal of patience and persistence. Publishing and selling content online is not a get rich overnight scheme, but it represents a long-term opportunity. You will need to consistently produce high-quality content over a long time. Over time you can make a good income by doing this. For this to happen, you will have to develop some work habits that will allow you to become self-motivated and to have consistent production. Unfortunately, the odds are against you if you hope to only write one book and expect that it will become a best-seller. Rather, the formula for success is to regularly produce content and build an audience over a period of time. This will require you to be able to produce content on a regular predictable basis.

1.2 Why Become a Writer?

"Of course it's why you want to become a writer - because you have the liberty to do that, but once you have the liberty you also have the obligation to do it."
Tobias Wolff

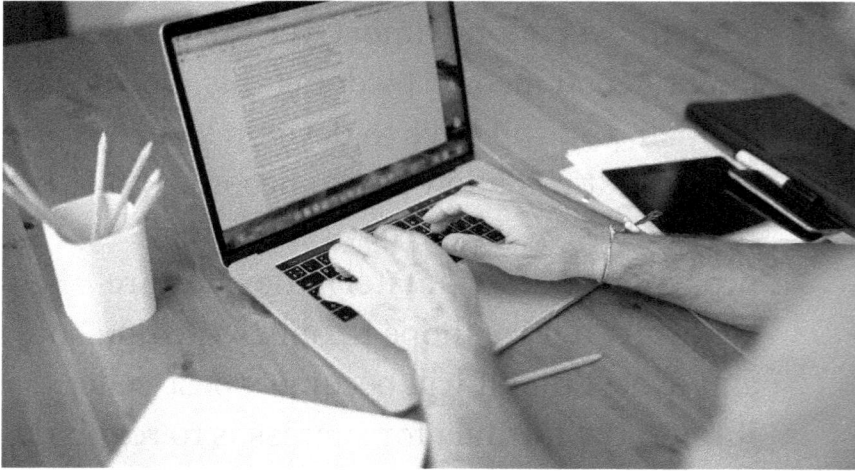

If your only goal is to make money writing, then I think you'll probably be disappointed by this opportunity. To make money, being an author will require you to put out a large amount of high-quality material over a very long period of time. Writing is a slow-growth, long-term opportunity. Although it is possible to write one book that becomes a runaway bestseller, it is highly improbable. The better bet is to develop a set of habits where you can put out a regular amount of material and grow your income incrementally.

Although it will take you some time to build up a steady monthly income, once you have established this, it is truly passive income. Once you have published a book, there is not much work to do beyond marketing the book to receive monthly royalties. In many cases, these books have very long life spans and will produce income for years to come.

There are a considerable number of benefits about becoming an author beyond just the financial aspects of the job. One of the truly amazing things about being an author is that you could literally change the world. The work that you produce can change somebody's life for the better and dramatically improve somebody's quality-of-life. The ability to do this is an incredible benefit that most jobs can not provide.

In addition to the life-changing aspects of becoming an author, it is an incredibly flexible job. In fact, I am writing this chapter on my cell phone, dictating it while I am sitting out on my back porch. Not many jobs can offer you the flexibility in terms of location that being an author can provide.

You will need only a minimal amount of equipment to get started and you probably already have what you need in terms of a PC and an internet connection. In fact, you may not even need an internet connection all the time. You only need one if you're using online tools to archive your work or to upload to publishing sites.

One of the other great benefits of being an author is that you can write whatever you want. Although there will be certain limitations about the type of content you can publish on some sites; you can literally write whatever is of interest to you. This process is much different than a corporate environment where you have to get the company approval

for the type of content you are producing. Also, you don't have to write the style that your boss wants you to write. Being a self-published author is incredibly flexible, and you can put your unique spin on whatever topic you choose to write about.
Self-publishing is a fantastic opportunity, and the tools that are currently available for publishing content has never been easier.

Another advantage of being an author is that it will establish you as an authority figure in whatever area you choose to concentrate. Even if you are a self-published author, there's a certain amount of prestige about having a volume of published material and a following. It allows you to speak from a position of authority when you go to meetings or presentations where you can cite references of your publications, and point to an online presence where you have a substantial following.

The road to becoming a self-published author is relatively difficult as it requires a large amount of self-discipline. You don't have a boss dictating to you what your projects look like or your next deadline; it is entirely up to you to establish your pace of work and your deadlines. For many people, this is exceptionally difficult to do, and unfortunately, they begin to procrastinate and never finish their first book.

However, I have found that with practice, you can routinely produce content even under unusual circumstances. I have learned to write under many different conditions. Recently I started using the cell phone to dictate content. Using this technique, I can create content from literally anywhere in the world. I can knock out a chapter in places where I would've been completely unproductive in the past. I recently was able to get some work done while stuck in an airport. I would have struggled to do this in the past. I've also now found with this freedom I enjoy mixing up the venues where I write so that it does not become an arduous task.

Another terrific advantage about being an author is that you can start part-time with very little investment and begin to ramp up as your sales grow and you expand your content reach beyond just printed books. It is a unique opportunity and allows you to test the waters without dumping a considerable amount of money into start-up costs. Today's book vendors such as Amazon print books on-demand, are designed so you do not need to invest in a large inventory of printed material to bet on future sales.

By having access to easy print on demand services such as Amazon Kindle, there is literally a zero barrier of entry for new authors to enter the market. Perhaps the most challenging obstacle venturing out is training, and that's why I wrote this book — to help new authors overcome some of the difficulties of publishing their first book.

Although companies like Amazon have significantly streamlined the process of self-publishing books, the first time through can be a bit daunting and frustrating. However, after this, you will find that Amazon Kindle is an amazing platform and allows you to get started with a minimal amount of effort.

Even with a large number of existing authors that are competing for business on platforms such as Amazon, now is the perfect time to start your publishing business. The demand has never been higher, and the entry barriers have never been lower.

Although it may take you multiple years to ramp up your business to have a substantial recurring monthly income, I have found that this is one of the most rewarding occupations I've ever been involved in. It allows me to express my ideas freely and publish from whatever venue I choose to a worldwide audience. Self-publishing is an incredible opportunity, and one of the things I genuinely enjoy.

I hope that this book now only gives you some tips to make your journey a bit easier, but also inspires you in some way and help you deal with the ups and downs of being an author. Good luck in your self publishing career!

1.3 LAB Write Down Your Goals For Becoming a Writer

In this exercise you will write out your high level goals for becoming a writer. Although you should think about financial goals, try and not make it your first focus for this exercise. You should include some production goals for yourself that are focused around producing high quality content on a predictable schedule. You can also include some goals that will help you build the infrastructure to support your book writing business. I have included some sample goals to help get you started.

Sample Book Writing Goals

Short Term Goals
- Read books on self publishing.
- Find a reliable and reasonably priced book cover creator to use on my first project.
- Complete my first draft of My First Book and publish it to Amazon Kindle by June 2020.
- Publish my second book by Dec 2021.
- Publish both books to IngramSpark to establish a second income stream.

Mid Term Term Goals
- Complete my first book series by 2022.
- Expand monthly income from book publishing to $1500 to cover help with monthly expenses.
- Build a dedicated home office to use as my writing studio.
- Start a YouTube channel in 2022 to help market books and provide supporting material to books.

- Sign up for training on how to shoot YouTube videos.

Long Term Goals

- Reach a production capacity to produce 5 books annually.
- Establish a monthly income that will allow me to work part-time within 5 years and supplement my retirement income.
- Expand selected books to video courses on Udemy.
- Leverage books and online material to book speaking engagements.

1.4 How Do You Get Started?

"The secret of getting ahead is getting started. The secret of getting started is breaking your complex overwhelming tasks into small manageable tasks, and starting on the first one."
Mark Twain

It all starts with an idea. If you are reading this book then you probably have an idea of where you would like to start. Many people that are starting or thinking of starting have an idea or an inspiration for their first book. If this is true for you, I would go ahead and get some words and outlines on paper and get the ball rolling. The more you flush out the book, the better chance you have to get input and feedback from other people. I have found that it is much easier to share an outline with one of my colleagues to get feedback than it is to verbally describe the book to them and then ask them what they think. The more concrete you can make it, the better chance you have of getting early feedback.

The other thing you might consider at this point is to begin to plan a roadmap of content that will follow your first book. Since the best way to make a steady income at publishing is to produce content regularly, planning your roadmap early on will help with this process. This roadmap does not need to be complete, or planned out many years in advance, but it will begin the process for you to accumulate some topics that will feed your production cycle. This roadmap will undoubtedly change over time based on your impressions, changing interests, and market conditions. Starting this process early on will help you transition more easily into your next project. I wish I had done more work on this as I started my journey; it would have helped me get over the dry spell I had early in my writing journey.

Another tip I have for you in starting is to pick a more manageable project that will help you get the ball rolling. If you can try and scale your first project to be a doable size that will be manageable for you to produce. Try and scale your first book to be manageable in size and complexity, that allows you to get a successful project out the door and establish some momentum for yourself. For example, if you want to write a book on a complex technical topic, perhaps you can write a sorted book that is an introduction to the topic that will be much shorter and easier to write and research. By doing this, you will have several advantages. The first is that you will see how well your audience receives your introductory content. The second is—you will begin to build an audience and have a potential market for the sequel. This technique may not always be possible, but if you can take advantage of this, it will make it easier for your first book launch.

Once you get started with your first book, make a schedule, and stick to it. Sit down and rough out a schedule chapter by chapter. Don't make your plan too unrealistic, but also don't make it too long of a period. I have found that I can now produce a book in a relatively short time if I am organized and motivated. I do a much better job of writing a complete book over a period of weeks than I do over a long period of months and months. This schedule may not be possible for the type of material you are producing, but the shorter the timeline you can make, the better.

Many people think of writing a book as a bucket list type of item and a real-life achievement. I agree with this, as it is a process that requires a great deal of self-motivation. However, if you are pursuing this as more than a hobby, you will want to treat your first book as more of a milestone on your way to bigger and better things. Take the time to celebrate your first release, but don't build it into more than it is. You will learn a ton about writing and your work ethic after this milestone. You can use this knowledge to improve your processes and techniques for your next project.

Becoming an author is no different than any other skill or profession. It requires skill, practice, and repetition to improve at your craft. It is highly unlikely your first book will be a masterpiece, so try and limit your expectations. However, it is highly likely if you follow some of my tips in this book that your first book will be a quality product that you can produce on a reasonable schedule. It will also set the stage for you to create a series of books that will lead you to build a growing following and successful results.

In summary, the hardest thing in any project many times is just to get started. If you have been thinking about writing a book, go ahead and start with the simplest thing you can do. Write a title down and then flush out an outline. On your lunch hour, find a quiet place, and plan out your schedule and get going. Don't overthink it. Get started today!

1.5 What Should I Write About?

"If everyone is thinking alike, then no one is thinking."
Benjamin Franklin

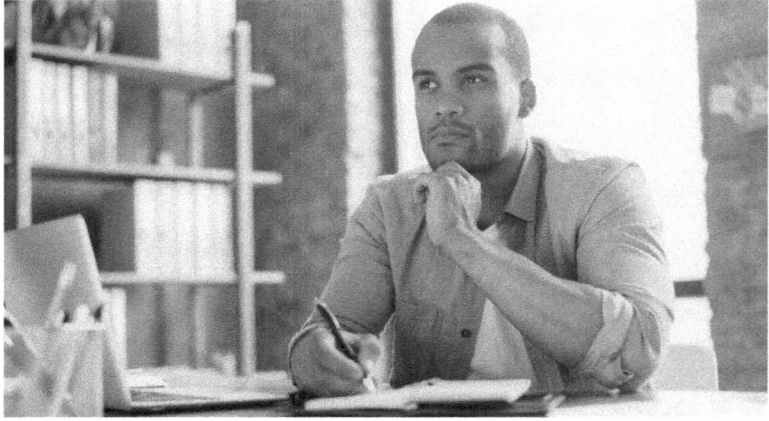

Probably the first thing you should think about is what are you interested in writing? Also, determine what your comfort zone is and what you are comfortable in writing. I come from a scientific and engineering background, so I am comfortable writing about theoretical topics and how-to books. I would have a difficult time creating a novel for the first time, although I would like to attempt this at some point in my career! Develop a list of potential topics and build a list of these. I have included a sample from my book list in the appendix of this book.

The next step in the process is to consider the sales potential of your book. Although I warned you about the primary goal of making money earlier in this book, it is a reality that you will want to grow your income from your publications. At this point, you can begin to research your topic to see how popular that might be.

The first step for me is to go out to Amazon and begin to search for my topic and look for similar books. When you find a book, look up the Amazon Best Seller Rank, as this statistic shows the rank in sales of all books on Amazon. You can then use this metric to determine approximately how many books per day that title is selling by plugging this number into the following free tool:

https://kindlepreneur.com/amazon-kdp-sales-rank-calculator/

Once you get this number, you can calculate the estimated monthly sales (multiply this number by 30) and then calculate the estimated monthly profit. (I use a figure of $3.00 per book based on the types of books I write). Let's look at an example of this.

In this example, I will look at books about the C# programming language. This area is one of the technologies that I am interested in and have written about in several articles and books. The following book was the top-rated book that I found.

amazon prime

All ∨ C#

Deliver to Eric Gahanna 43230

Fresh ∨ Whole Foods Black Friday Deals Best Sellers Sell Prime Video Eric's Amazon.com Help Browsing History ∨ Find a Gift Buy Again Gift Cards

Books Advanced Search New Releases Best Sellers & More Children's Books Textbooks Textbook Rentals Sell Us Your Books Best Books of the Month

Shop Black Friday deals starting today

‹ Back to results

C#: Programming Basics for Absolute Beginners and millions of other books are available for Amazon Kindle. Learn more

C#: Programming Basics for Absolute Beginners (Step-By-Step C#) 2nd Edition

by Nathan Clark ∨ (Author)

★★★★☆ ∨ 53 ratings

Look inside ↓

Save

| Kindle | Paperback | Other Sellers |
| from $2.99 | $15.56 | See all 3 versions |

Buy new

FREE delivery: Monday Details

⊘ Deliver to Eric - Gahanna 43230
Available to ship in 1-2 days.
Ships from and sold by Amazon.com.
√prime

√prime $15.56

3 New from $15.56

Qty 1 ∨

Add to Cart

Buy Now

More Buying Choices

3 New from $15.56   3 Used from $15.56

6 used & new from $15.56

See All Buying Options

★ C# Made Easy – a Step-by-Step Guide for Beginners ★

Learning a programming language can seem like a daunting task. You may have looked at coding in the past, and felt it was too complicated and confusing. This comprehensive beginner's guide will take you step by step through learning one of the best programming languages out there. In a matter of no time, you will be writing code like a professional!
Read more

ISBN-13: 978-1975745080
ISBN-10: 1975745086
Why is ISBN important? ∨

Have one to sell?   Sell on Amazon

Add to List ∨

Share ☑ 📘 💬 ⊕

□ Report incorrect product information.

See the Best Books of 2019
Browse the Amazon editors' picks for the Best Books of 2019, featuring our favorite reads in more than a dozen categories.

If I scroll down, I can find the Amazon Best Seller Rank for this book.

Product details

Series: Step-By-Step C# (Book 1)
Paperback: 135 pages
Publisher: CreateSpace Independent Publishing Platform; 2 edition (August 24, 2017)
Language: English
ISBN-10: 1975745086
ISBN-13: 978-1975745080
Product Dimensions: 6 x 0.3 x 9 inches
Shipping Weight: 7.7 ounces (View shipping rates and policies)
Average Customer Review: ☆ ☆ ☆ ☆ ☆ ˅ 53 customer reviews
Amazon Best Sellers Rank: #568,171 in Books (See Top 100 in Books)
    #181 in C# Programming (Books)
    #309 in Microsoft C & C++ Windows Programming
    #1208 in Introductory & Beginning Programming

You will note the ABSR us 568,171 for this book. If I plug this into the Kindle Best Seller Calculator, I get the following result.

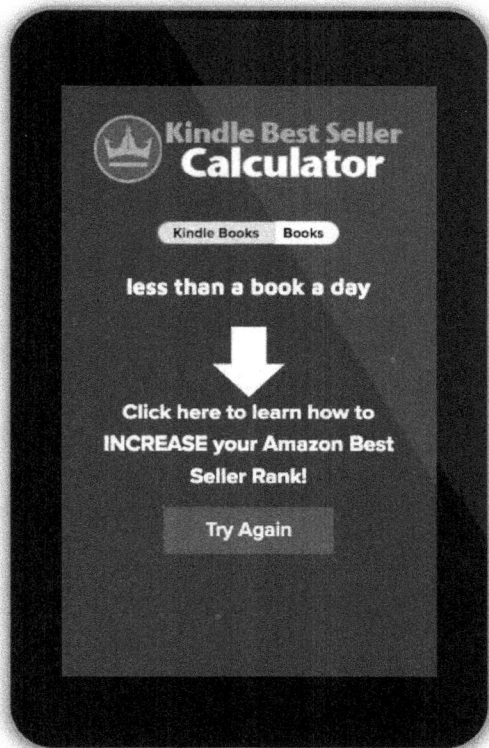

I get the result of less than one book per day. For our purposes, I will assign the value of .25 book per day based on that the estimated monthly revenue would be.

0.25 book per day *
30 days per month *
$3.00 profit per book =
$22.30 profit per month

Let's look at another topic and compare it. For this example, I will look at another programming language, Python, and do the same calculation.

The listing below is the first book that I found on the list and will now check out the Amazon Best Seller rank for this book.

Product details

Series: Python Programming (Book 1)
Paperback: 187 pages
Publisher: Independently published (October 26, 2019)
Language: English
ISBN-10: 1697526373
ISBN-13: 978-1697526370
Product Dimensions: 5.2 x 0.4 x 8 inches
Shipping Weight: 8 ounces (View shipping rates and policies)
Average Customer Review: ☆☆☆☆☆ ˅ 44 customer reviews
Amazon Best Sellers Rank: #39,230 in Books (See Top 100 in Books)
   #23 in Object-Oriented Design
   #68 in Software Development (Books)
   #15 in Computer Systems Analysis & Design (Books)

For this book, I got a much lower ABSR rank of 39,230. I will now go and put this value into the Kindle Best Seller Calculator.

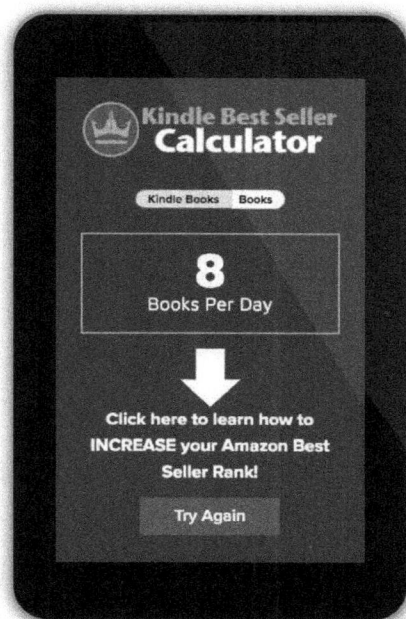

I get the result of less than 8 books per day. Based on that, the estimated monthly revenue would be.

8 book per day *
30 days per month *
$3.00 profit per book =
$720 profit per month

You will want to do this for several books in the category to get a better idea and then average these out. But based on my preliminary analysis, it looks like there is quite a bit more interest in Python programming books over C# by quite a bit.

After you have completed this exercise, you can then judge all of the topics you are interested in and then decide on the balance of what topics you are interested in versus those that might be the most profitable. I should note that this is not an exact science but can give you an idea about what people are buying on Amazon and where you might be able to carve out a profitable niche.

I will also note that the more popular the area is many times it is more competitive since there may be many books already in this category. The ideal pointis to find a reasonably popular niche that is not overly competitive. I would not obsess over this for your first book, however. Your focus should be on establishing

a regular repeatable that produces
high-quality content.

1.6 LAB Develop Your Book Idea List

In this exercise you will begin to develop your book idea list. This will be an ongoing tool for you to use as you progress in your writing career. I have found this is a good thing for me to work on while I am on long airplane flights. You can use this as sort of a notebook that you can work on or jot things down in when the ideas come to you. I like to do the same thing with book covers to help me with the creative process.

The premise behind this is that you will not be a one hit wonder, but will develop a series of books that allow you to build a portfolio of material that will provide you a steady income and a base to build your build your publishing business. Even if some of the ideas seem far flung or a stretch to you, put them on the list for future consideration. You never know how your interest or the market conditions might change. When you have an idea, simply add new titles to the list when the ideas come to you. Later you can update the sales potential of each book and then you can sort the list on this and other criteria.

I have provided a short sample list below to help get you started. You can also go to my website and download and Excel spreadsheet that is a template for this. You can get this spreadsheet here: hpps://detsinlearning.com/toolkit

Book Title	Subtitle	Research Required	Sales Potential
Introduction to Java	N/A	Low - based on previous teaching lessons	Med - Lot of books in this space
Principles of Software Engineering	N/A	Med - based on my software mgmt experience	Low - specialized topic
Introduction to Excel	How to Build Practical Solutions with Excel 2019	Low - based on classroom labs and experience	High - although there is a lot of competition in this area
How to Become a Better You	A Series of Life Lessons	Low - based on life lessons from my mentors	Low - High competition in this area. Will write for personal reasons.

2.0 Writing Your Book

2.1 The Book Writing Process

"Either write something worth reading or do something worth writing."
Benjamin Franklin

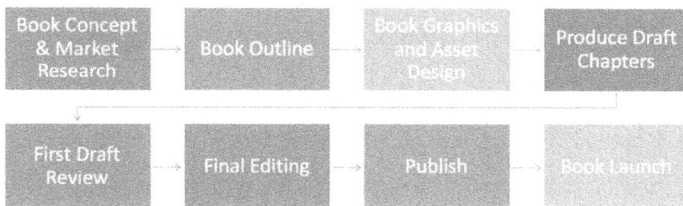

I will tell you that writing is not a one size fits all process. We each process things in a different way. So with this in mind, I am going to describe the process that I use, and how it helps to keep me on track. You may find that some variation of this process works better for you, but the necessary steps will still be the same. I am a very visual person and many of the process elements that I have migrated to in my book production process helps me to visualize things each step of the way. You may not be as visually oriented, so take that into account when you're thinking about these steps for your process. You might want to change them slightly to make them work better for you.

Also, before you begin to tailor the process for your own projects, I suggest you use this process as is for your first book project. Getting your first book out the door and into production is very challenging the first time though. It will become easier over time and will become much more straight forward after your first book. This is one of those tasks that you will only learn by doing it yourself.

1 Book Concept & Market Research

The first step in the process for me is to develop a book concept that has the general idea of what I want to write. I generally have a lot of concepts floating around, and I keep a log of these. When I think of them, I can write them into my book so that I don't lose the idea later. Once a concept becomes clear for me, I will go out and do a little bit of research to see what other types of similar material may already be out on the Internet.

I will check on sources such as YouTube, Amazon.com, and other book sites to see if there is a similar book that is already in production. If I find similar books on Amazon, I will then look at their sales rank, ratings, and use the Kindle Best Seller Calculator. Also, I will read select reviews to see what the readers are saying about the book. I will also look at the contents of these books or videos. I want to make sure that I have a bit of a different spin than what is already out there. If there are a lot of topics in a particular area, I won't look at it as a particularly negative influence. However, I will know that I will have to do a particularly good job in developing the material to be competitive in that area.

Also, during this phase, I will begin to think about a book cover and start drafting some ideas to see how they might look. I use an internet-based tool called Canva to do some mockups and see how they might seem. I will generally try to draw some graphic images that I can use as a centerpiece for my cover. I may also look for a stock photo image on a site such as Shutterstock to see if there is something already out there that will fit my needs. In some cases, I have used photographs that I have taken myself that just happened to work out for a book cover. Using my photos is generally not the case and is somewhat difficult for me to do since I am not a great photographer.

It may seem a little early in the process to develop a book cover, but it helps me crystallize the concept in my mind for the entire book. The more polished and final the book cover is, I have found that it helps me in developing the remainder of the material for the book. Since I have that image in my head that I have drawn with a detailed book cover, I can then fully expand all of the topics in the book. I know the process that I have developed is a very top-down process. It is likely a process that I learned while I was in engineering school and have worked this way on for many engineering and software development projects throughout my career.

For those of you that like to work bottom up, you could move this part of the process later on to gather several chapters that you might write independently to stitch together a book. I have not worked in this manner to construct a book, so I can't comment on its effectiveness. I realize many people don't work from the top down and like to work from a lot of little details and gather those into a final project. The other thing I like about the top-down process is if I stop a project I at least have a concept that's pretty fully baked. In this way, I can come back to it anytime and finish the process when I have more time.

I have tried making my own book covers for my early books, but they truly lacked the pop I was looking for. I then looked at a number of services that offer booked cover design services. Most of those that I found were very expensive and I did not want to invest that much money into the process. I then began looking into services on Fiverr (http://www.fiverr.com) and found a number of really good services that were very reasonably priced. I now use Pro eBook Covers for all of my book covers. You can check out the service at the following link:

https://www.fiverr.com/pro_ebookcovers?source=order_page_summary_seller_link

I can't recommend this service enough for doing book covers. She really does a fantastic job. In fact, just this tip alone is worth the purchase price of this book. The service that they provide is much better than services that cost hundreds of dollars more. You will have to provide a royalty-free image for this service to use for the book cover as well as links to two books that they can use as a basis for the image cover design. These are both relatively easy to find. There are a number of free royalty-free image sites and a number of paid sites such as iStock and Getty images.There is a list of the URLs for all of the services I have had good experience with, and that I can recommend, in the Appendix of this book.

2 Book Outline

After you have completed your book concept, now it's time to prepare the outline for your book. Although you may be tempted to jump in and start writing right away, I have found by spending some time working on a detailed outline, it helps me organize my thoughts and makes the book a complete work. The more and more I write, the more helpful I am finding the outlining process can be. The more detailed you are in this process, the more likely it will pay dividends later when you're cranking out each chapter.

Try thinking from the top down and develop some goals for your book. Make sure that your outline accomplishes those goals and objectives which you have outlined in your list of goals. Later on in this chapter you will find an exercise in producing an outline for your book. I have also included an example of an outline of a book that I'm currently producing that can use it as an example.

After you have completed your book outline, it's a good idea to send the draft out to people who will be willing to peer-review your book. If you package a cover along with an outline and ask them if they're interested in helping you review your book, they will have a good idea of the content and if they are a good fit for the review. I found this to be a much better situation to send out material for the review rather than just asking them about a subject and if they might be interested. Having concrete material to send out at this point really stimulates a lot of interest from potential reviewers. Most of the time they are very willing to help. Also, getting feedback on a cover, and an outline is a much simpler process than getting feedback on an entire manuscript. A reviewer will have to invest quite a bit of time reading the whole book.

In my experience, I have found that once I've completed the cover and have a solid outline, the rest of the process is much easier. Using these as a baseline, the remainder of the writing process is much more routine.

3 Book Chapters and Asset Design

You may think that his next step is not for you, but it's one that has really helped me in my book production process. At this point in the process, I like to pick out all of the graphics and visual information that I would like to use in my book and begin to lay them out where they will be in each chapter. Since I'm a very visual person, this process helps me think of all the content that I want to include in each section and what visual representations I would like to use to reinforce those concepts.

I write a lot of technical material, so this step is critical for me to include all of those visual details that I need to use to reinforce my points. If you are writing a novel or something that is not generally very graphical and nature, this may not be an essential step in the process for you. However, I have found, if you are producing any book, it is good to have some types of graphical images, particularly in the headings of chapters, to help make the reading experience a little more interesting. These graphics help the reader visualize the material that you are trying to present.

Again this is step entirely up to you, but I have found that once I've completed this process, it gives me a jumpstart as I'm producing each chapter. Over the years, I have also built up a library of graphical assets that I can use to help make this process go a little quicker. I also have an ongoing account with iStock images that I have a monthly quota of images that I can download for use in my books, videos and online courses. If you think you are going to produce this type of material, I highly recommend that you get an account with a stock image site so that you will have material that you can quickly pull to use in your content. I've included some additional information about some of these sites in the Appendix of this book.

4 Produce Draft Chapters

I have designed the preceding steps to set you up for success and producing the draft chapters of your book. I must admit this is the most challenging process for me to accomplish and is probably the most difficult for any author—to be able to produce draft chapters regularly and consistently. The other steps proceeding are pretty easy for me to get going; likewise, it's easy for me to stall in the chapter production process.

In this part of the process, you will write down content for all of the chapters in your book. It would be best if you worked on a sequential basis, starting with the first chapter, writing that material, and then moving on to the next chapter. At the end of this process, you may want to rewrite the introduction as you might have included material that you had not intended to in the beginning. At this point in the process, I usually put up a checklist of all the chapters I need to write on my whiteboard, and then I manually check these off as a write each chapter. This process may be an old-fashioned way of doing things, but I like the visualization that I can see my progress on my whiteboard. It also keeps the project in front of me, so I know when I'm slipping, and I can react accordingly. The key is to have a checklist to guide you and give you a way to gauge your progress.

I have found some tricks along the way that worked for me—that help me keep my production high in order to keep the process fresh. I am not a particularly good typist, and it has been my Achilles heel and producing content. Because of this, I have experimented with dictation so that I can speed up the production process and free up my mind while I'm writing. The dictation tools have gotten much better over time, and I now use the Voice Typing feature of Google Docs regularly. By using this, I can concentrate on the flow of the thoughts that I am trying to put into my material versus the exact sentence structure I need to use to be grammatically correct.

I use a two-step process; first, I will draft out a section or subsection of a chapter using a brief outline that I have jotted down on a piece of paper as my input and will dictate that particular section of the book. In the second step, I work through and clean up those sections that I've dictated using Grammarly Professional. By doing this, I no longer dread sitting down writing out chapter after chapter of content, and it allows me to do this reasonably quickly. The downside of this methodology is it really depends on having an excellent editing process to catch any dictation errors. I have found that Grammarly is an excellent tool for the first pass of editing. I rely on the final editing process to clear up any remaining issues.

You will have to experiment with your own process to see what works for you. I found that dictation has taken a lot of the drudgery out of writing for me and has exponentially increased my content production rate.

5 First Draft Review

Once you have finished writing all the material, you have reached a significant milestone by completing your first draft. Many times this is a difficult obstacle to overcome in order to achieve this milestone. At this point, you will want to make sure that you are comfortable with the material that you've produced, and it's ready to go out for final editing.

I recommend going to a quiet place and reread the book from beginning to end and use this as an opportunity to find any glaring errors and rerun any tools that you used, such as Grammarly. You will want to make sure that you have not left out any significant material, and you might use this as an opportunity to go back to your outline and recheck all the thoughts to make sure you have included them in the text.

Do not use this part of the process to obsess over your book. The main goal of this process is to make sure that the book is ready to go for final editing and that you haven't missed any significant portions. It's easy at this point to second-guess any of the work you have done. However, it would help if you remained confident that you have followed a process, and you thought through all the details as you're producing the book.

It's a good idea to let some time pass after you finish writing and before you do the final review. There is a strange phenomenon that seems to work here. If you review the content you have just completed, it appears to be okay. If you wait a few days or weeks, you will likely find obvious mistakes. I'm not quite sure what causes this, but I found if you can at least wait a few days or maybe even a week, it's better for doing your final review. This waiting time, however, is dependent on the type of schedule that you have, and generally speaking, most projects are running late at this point due to the fact it takes longer to write out the chapters than you initially expected.

Once you've completed your final review, go ahead and send it out for final editing. It is also an excellent time to send it out to anyone you have contacted to be on your launch team for a peer review. The purpose of this is to get feedback on the material and not to help with grammatical editing. Explain to them that you are looking for feedback on the high-level content of the book and determine if the book is matching the goals that you set out when you develop the concept.

It's a good idea to contact some of these people when you're developing the book concept. Contact them once you have created the cover so you can get feedback not only on the idea but the book cover as well. You can also ask him at this point if they would be willing to help with your book project by reviewing the content. You might also ask them if they would be willing to post a review for your book once it goes live into production. Be sure to include that you want their honest opinion, and you are not looking for just rubber stamp reviews.

6 Final Editing

In this step, you will send out your book for final editing. The role of your editor is to clarify anything not clearly worded and also to correct any grammatical errors and spelling mistakes. If you have an editor that is also a subject matter expert, they can also comment on the content as well, but generally speaking, I rely on an editor just for the mechanical aspects of the book.

At this time, you can also send out your book for peer review. You are looking for different input from those doing peer reviews than you are from somebody who is strictly editing the text. You will want them to concentrate their efforts on the content and concepts presented in your book to make sure they are clear and consistent. This task is particularly important in technical works.

During this stage, I like to use the Google Docs sharing feature. It makes it much easier to share documents with other people and consolidate all of their comments in one place. Before using this, I was emailing documents all over the place and was having difficulty tracking the versions of documents that I had emailed out to various people.

Once you get all of the comments back, you can consolidate them and make any corrections to your book. After completing this, your book will be ready for publishing!

7 Publish

Now that you have completed the text, it's time to publish your book on your desired platform, such as Amazon Kindle. In this step, you will format your book to the target file format, such as an Epub file. You will also have to paginate your book to make sure that once you've converted to the desired format the pagination is correct. It can be somewhat tricky in ebooks since ebook readers can display ebooks on multiple platforms in multiple sizes and fonts. If this is your first time through, I recommend using the Kindle Create software to help you format your final version to upload to Amazon.

Amazon supports multiple input formats. I have standardized on Epub for my setup. I have found that this format is much more consistent for me, and most platforms other book platforms support this as input. As I mentioned earlier in the book, I use Apple Pages to create my Epub files.

At this point, you can also perform at your book as a print book as well. Amazon consolidated this function into Kindle about a year ago and retired the Create Space product. Surprisingly I found that my print books easily outsell my ebook, so don't overlook this opportunity to diversify your income stream.

You will have a couple of decisions to make regarding your print book. The first is the size of the book. This size is referred to by book publishers as the trim size. Amazon offers many different options for trim size, and you will have to decide what is right for you. The most popular trim size is six by nine inches. To publish your print book, you will have to upload the manuscript, the book cover, and the book description.

After you have loaded your materials for your ebook and print book, Amazon has a preview feature that will allow you to page through your book and look for any mistakes.

When you have completed all of these steps, you can set pricing for your book. Amazon has some excellent documentation that will assist you in setting pricing for your book. For ebooks, there is a pricing wizard that you can click on, and it will try to determine the optimal price for you to set for your book. For you to get maximum revenue share on your eBook, you will have to set pricing between $2.99 and $9.99. You can set pricing higher than this on Amazon, but you will get a lower percentage revenue share. I have included links to the Amazon documentation in the appendix of this book.

Once you're satisfied with the results, you then click the publish button. This action triggers a review process from Amazon but generally only takes 24 hours to complete. After that, your book is live and available on Amazon

Marketplace! Congratulations, you have published your book! There are detailed step by step instructions on how to publish both eBooks and print books on Amazon Kindle later in this book.

8 Book Launch

Now that your book is live, there are a couple of items to do related to the book launch. The first activity I generally do is to order some author copies of the book. Most book publishers offer author copies of the book at a discount so that you can buy them and have them shipped directly to you. I generally do this right away so that I can proofread the book for any mistakes. I also keep a few copies that I can give away to colleagues that might be interested in reading the book.

If you have a launch team in place, now is the time to notify them that your book is live and in production. If they care to place a review for your book, that is a significant first step to get some early feedback on the text and to get the ball rolling with at least some initial reviews. I should stress at this point that Amazon's terms and conditions strictly forbid paying people to do reviews on your book. You will have to get people that will be willing to help you by reviewing your book honestly and openly. Getting book reviews is perhaps one of the most challenging tasks of publishing a book, especially early on when you first get started. Later, after you've developed a more substantial following, this gets a little bit easier. Don't try and shortcut this process by paying for reviews. I think you will regret it later.

At this time, you can also begin to post material on social media outlets such as Facebook and Linkedin announcing your go live for your book. You can start marketing activities at this point as well, such as ads on amazon.com or other platforms such as YouTube.

If you find mistakes in the original author copies of your book, now's the time to correct them quickly. Fortunately, on Amazon, the book revision process is straightforward, and there is no charge for this. This process is not free on other book publishing platforms such as Ingram Spark, where there is a $49 charge for each book revision.

9 Summary

In this chapter, I have outlined several tasks in a process that you can use to produce a book. This process may seem overwhelming at first, but I think if you work from the top down and chip away at it a little bit at a time, you will be successful. Producing your first book is a difficult task and a hurdle that you have to get over mentally more than anything to complete the job. Break down your projects into many small tasks that you can work on a little bit every day. By doing this, you can produce an enormous amount of work in a very short time.

Concerning time, I highly recommend that you develop a schedule for your book production to keep you on task. This schedule should be compact enough that you can work on it a little bit at a time but not so short that you feel like you are rushing through the process. I think to be successful you should make your schedule as quickly as possible. This schedule should be in terms of weeks or months and not be thinking of six months to a year for book production. I would start the smallest book that you can produce to at least get some experience under your belt before going after more significant works. Larger projects are much longer and more challenging projects to manage. If you are like 99% of the authors in the world, you won't be a one-hit wonder. You will make your mark by regularly producing quality content and doing it on a routine basis.

Don't go overboard with your planning. Keep it simple and make your schedule as simple as something you can put on a whiteboard or a piece of paper. The schedule could be something you could tape to a wall that you can look at from time to time. This visibility will help you make sure you are moving forward with your project. You will find over time that you can write a considerable amount of material. This progress is only possible if you have the right environment, you're motivated, and you feel like you're making progress.

2.2 How to Develop Your Book Cover

In this section, I will describe some of the things you can do to help you develop your book cover. This process is a creative process and requires a bit of trial and error. Because of this, it's not a step-by-step cookie-cutter process but several activities that you can undertake until you achieve some satisfactory results. I have included some tips that have helped me along the way.

The first thing you can do is sign up for an account at canva.com. Canva is a great graphics drawing package that's available online through a web browser. They have a free account that you can start with, and you can use it for all different types of graphic images, including book covers. They have some templates that are free to use, and I like to use these as ideas for generating a book cover once I have started researching a new title. There are some license restrictions around using these to publish your book, so be sure and read the fine print if you decide to use one of these templates.

The next suggestion I have is to go out to Amazon.com and look at similar book titles. Studying related titles is an excellent way to get ideas for your book cover. I'm not suggesting that you lift designs from someone, but you will definitely see book covers that you like and don't like, and some of the techniques that may be helpful in your design.

At this point, you can also begin looking at some stock images that you can use to include in your book cover (see the appendix of this book for several stock image sites). In addition to the paid stock image sites, there are also many free stock image sites that you can look at for ideas. Make sure you have checked the license agreement for the stock image and that it is suitable for publishing with your book. If you have any questions around this, contact the stock image provider and ask them questions around publishing your book.

By now, you should have a couple of rough drafts put together based on the material that you've researched. You can use a tool like PowerPoint or Canva to generate a couple of draft book cover concepts that you can send to your friends and family and see which ones they like. Sending these covers out for review is a very inexpensive way of doing an A/B test to see which cover is more effective.

After you've received feedback on your draft book covers and you selected a final design, you can finish it off yourself with a tool like Photoshop, or you can hire it out to a book design service. If you decide to do it yourself and use it to like Photoshop, you should be familiar with using that tool or something like it. I have tried this in the past and if not been very successful for me. I now hire out all my final designs. I have been delighted with the work that I've received back. Also, the designs that the service has produced are much better than the ones I did myself. However, if you have the talent and the training to use software such as Photoshop, you may be able to produce very good results on your own. The choice is entirely up to you.

Here is an example cover of a book that I'm currently working on. The first image is the concept image that I developed after looking at some stock photos and other similar books. I then sent this material to Pro ebook covers on Fiverr for them to finalize the design. The next image is the final design that came back from Pro eBook Covers and the one that I will use for production.

Introduction to Microsoft Excel

How to Build Practical Solutions with Excel 2019

Eric Frick

Rough Draft Book Cover Concept

INTRODUCTION TO
MICROSOFT
EXCEL

**How to Build Practical
Solutions with Excel 2019**

ERIC FRICK

Final Book Cover Design

2.3 LAB Develop Your Book Cover

In this exercise, you will develop the cover for your book based on some of the ideas that I've given you. At this point, you can go out to Canva and sign up for a free account to begin to draft out your design ideas for your book cover. I recommend that you come up with a couple of basic designs and then run them by your friends and family to see which ones they prefer. This exercise is a cheap way of doing A/B testing, and you can get input right away to see which designs are more effective.

If you would also like to invest a small amount of money at this point, you can contact the link that I've recommended for Pro-ebook Covers on Fiverr and develop a basic cover design for your book at this point. The baseline service for Pro-ebook Covers starts at $20, and I think you will get some excellent results. For your order to be processed, you have to provide a stock image to include in your cover that has rights that allow you to publish it on a royalty-free basis. You will also need to provide them with two links to books that are similar which they can use as a design reference.

I think this is a good point in the book to stop and complete this exercise. Once you have a book cover, you can use this as a visualization of your project and show it to some of your friends or co-workers. You can show them the idea that you're trying to produce. It will help you to have something to send to them so you can gather people that will help review your book and give you feedback on the material. I have discovered that if I have the cover at this point, it is a great conversation starter, and you will get feedback right away on the cover as well.

2.4 Book Writing Tools

"Technology is nothing. What's important is that you have a faith in people, that they're basically good and smart, and if you give them tools, they'll do wonderful things with them."
Steve Jobs

There are a number of really great tools on the market to help out writing a book. The problem is there are so many tools that it can be difficult to decide which ones to use. I have found that many of these tools were too complicated for me to use unless I used them on a daily basis and became very proficient with them. If you are already comfortable using some of the tools I mention below it will make your decision process much easier.

Microsoft Word

Microsoft Word is the de facto standard for word processing. Most corporate environments use Word throughout the world, and Microsoft Office dominates the desktop environment. Word is a very capable tool. It has tons of features and most publishing environments like Amazon Kindle support input files that use Word format. There are also templates available to assist with formatting and other features that can help you in publishing your book. Also, if you're going to share your writing with other people, the odds are that they have access to Word, particularly editors.

When I first started writing, I used Word as my primary production tool. I was very familiar with it and had used it in the business world for several years. After some time, I became frustrated with Word and had difficulty in producing consistency with large documents. My target format for publishing ebooks to Amazon is epub. I found that I was spending a lot of time trying to convert Word documents to the epub format, and I ended up wasting a lot of time and effort.

Also, I was emailing copies of documents to multiple people during the review process and began having trouble keeping track of att of the changes. Eventually, I moved on to other tools for my primary publishing tool and manuscript management. I did not try the online version of Word that is a part of Office 365. I know that this version has much better features for sharing documents with others and managing revisions.

In summary, Word is a great tool, and you can use it for almost anything, but for me, however, I have found other tools that better meet my needs.

Scrivener

Scrivener is a very popular tool within the writing community. It is specially designed for writers and has a lot of beneficial features that make it well suited for many writers. There is a version that is available for both the PC and the Mac. There is an active community that supports the software. Scrivener is also reasonably priced in the company supports the product with regular new releases. I've included a link for Scrivener in the appendix of this book.

The software has built-in features that support collecting and storing research for your book within a project. It also includes features for tracking progress on your project, including target word counts for your book. The navigation and the setup for moving around your book are ideal, and it helps with the creative process. Also, the software includes a type of idea board that you can tack up cards for ideas and relate them to chapters in your book. If you are a playwright, there are features integrated into the software to help you manage your play. It includes features to help with the personas of your characters and so on.

The software also supports some very robust export capabilities. It will export to almost any format that you can imagine, and it has a powerful and flexible way that you can format to multiple different outputs and have stored settings for each of those.

The one feature that is glaringly missing is the spelling and grammar checker in this software is horrible. You will have to integrate with a tool like Grammarly or some other tool to assist you with this. I have a love-hate relationship with this tool. I really like it for the creative process, but I found, in the end, I was spending too much time working on the formatting and was better off using a more straightforward tool to draft my books.

There is a trial version of this software that is available on the Literature & Latte site that you can use for 30 days. Even though I decided to move in a different direction, I still occasionally consider going back to the software due to its innovative design. You might consider giving this software a try and see if it meets your needs.

Google Docs

Google released Google Docs in 2009. The software originated from a company that was acquired by Google. After the original release of Google Docs, the company continued to enhance the product with a series of regular releases that increased its capabilities. Eventually, they released a commercial version of their online office products in a product called G Suite. G Suite is a competitor of Office 365 and includes a spreadsheet program and a presentation program as well as other tools.

I currently use Google Docs for almost all of my writing. I really like the simplicity of the product and also like the fact that it integrates with Google Drive, where I store all of my manuscripts. I also like the fact that Google easily integrates with Grammarly which I use regularly. Even if you do not decide to use Google Docs, I highly recommend that you periodically backup all of your manuscripts to a reliable place such as Google Drive or

Microsoft OneDrive. Writing books is hard enough without losing work.

Another feature that has drawn me to Google Docs is the ease in which you can share documents with other people. Although you can also do this with Microsoft Office 365, the Google version of this seems to be incredibly flexible. In fact, multiple people can edit the document at the same time, and you can see everyone's changes in real-time. It's incredibly useful if you're collaborating with other people on a document.

I also utilize several add-ons and help me with some specialized tasks that I need to perform in Google Docs. One of them is including computer code listings within my documents. I have found an add on called Code Blocks that is an excellent tool for this. Another tool that I use is for formatting tables (Table Formatter) that's very similar to the table formatting capability in Microsoft Word. There are literally hundreds of add-ons for Google Docs, and they extend the base functionality and help you do all sorts of things that are incredibly useful.

Once I have completed my manuscripts, I export my documents from Google Docs, usually in Microsoft Word format. From there, I use other tools to do the final formatting before I publish my books to either Amazon.com or Ingram Spark. Initially, I was looking for one tool to do all of these tasks and used Scrivener for this, but I found the process was too complicated to support. I was spending too much time messing with the tool and not enough time writing.

Apple Pages

Apple Pages is a word processor that comes with the Mac operating system. It is a complete word processor, and it can import from formats such as Microsoft Word and others. It also has the advantage that you can use it on a Mac, iPhone, or iPad. If you use Apple Pages on an iPad, there are some advantages that you can use a stylus to help edit your documents.

One of the more unusual features about Apple Pages is that you can also use it as a teleprompter. This feature can come in handy if you're going to create audiobooks from your scripts, and you want to record them or if you're going to produce videos from your material for something like YouTube. The software also works well in conjunction with Apple keynote, which is their presentation software and Apple Numbers, which is the spreadsheet that comes with the Mac.

The main feature that I use Apple Pages for is it has an excellent capability to export documents in the ePub format. I use this as my go-to format for Kindle eBooks. My process is straightforward. I export my manuscripts from Google Docs into and then import them into Apple Pages to produce my ebooks. Although this seems like a convoluted process, it has worked well for me and has saved me a lot of time.

Google Docs has the same feature that can export documents as epubs, but they don't look correct when you upload them into Amazon. I've also tried formatting Epub files with other tools as well, but I found the Apple Pages method to be the most straightforward and most reliable. I'm hoping to find a better process over time, but for now, this is the process that I use and is very straightforward for me. Since I create almost all of my material on a Mac, using Apple Pages has not presented any problems for me.

iBooks Author

Another tool worth mentioning on the Mac is iBooks author. The software also comes with the Mac OS and is specifically designed to produce eBooks on the Mac.

iBooks Author has some incredibly beautiful built-in templates, and it's straightforward to use. I was very excited to use the software only to be disappointed that Apple specifically designed it to be used with iBooks. It supports both templates for ebooks and print books; however, the output options are very limited, and the more I use it, the more frustrated I become.

I tried several times in vain to use it with Amazon Kindle, but in the end, it was too difficult to convert the files. Perhaps Apple will add features to the software to allow publishing to other formats in the future. As of the time of this writing, I did not see these features in the software.

If you intend to produce content solely for iBooks, this is a great tool to use, and you will create a fantastic looking ebook using this for iBooks. I also ran into some problems publishing to iBooks as a lot of my books reference links in Amazon.com. iBooks does not allow any links to Amazon in books that they publish as this is a competitor about Apple. For that reason, I do not publish any books to iBooks at the current time. That stance seems to be a bit extreme to me and stifles free expression.

Kindle Create

Amazon offers a tool called Kindle create. Amazon specifically designed the software to produce books for Amazon Kindle. This software is free to download, and there are versions for both PC and a Mac. The software is straightforward to use and can import Microsoft Word files and PDF files.

There are two versions of projects that you can produce with Kindle Create. The first is a reflowable format, and a source for that is Microsoft Word document. This type of ebook is the typical ebook that you would produce for Amazon Kindle. If you're creating a textbook or a book with lots of graphics with complex formatting requirements, you can produce a fixed ebook, and the input for that is a PDF file. There are some limitations if you produce this type of book on the devices that you are able to read the publication with, but if you're producing a textbook, this is the way to go to import a PDF file into Kindle create.

The software claims to have the capability to publish both ebooks and print books from a single source. I have tried this recently and was unsuccessful with the print book, but this is an incredibly useful tool to format books for publishing on Amazon Kindle.

One of the other features I like about this software is that it automatically creates a table of contents for your ebook. I have had lots of problems doing this, trying to upload other formats such as Microsoft Word Documents to Amazon Kindle.

If you are going to publish to Amazon Kindle using the software, particularly for your first time publishing a book will be incredibly useful and save you a lot of time and frustration.

Microsoft PowerPoint

The next tool I will cover is Microsoft PowerPoint. PowerPoint is not really a book writing tool, but you can use it to create short ebooks or brochures. It does an excellent tool to help create promotional material for your books, such as short handouts, summaries, or posters.

Powerpoint is also handy for creating illustrations and diagrams. Since I create mostly technical books, I use PowerPoint extensively for creating flowcharts, illustrative diagrams, timelines, and other technical documentation. You can export these illustrations from PowerPoint into JPEG or PNG images that you can include in your book. PowerPoint also has a smart art capability that makes producing nice-looking charts very simple.

Make sure when you export these images that they have a resolution of at least 300 DPI for use with print books. This resolution is suitable for print books on Amazon or providers such as Ingram Spark. This resolution is also applicable to any screenshots that you do for inclusion in your book. To make sure that screenshots have a suitable resolution, make sure you make your screenshots on a 4k monitor that will support this type of resolution.

I've also found it very useful to keep a directory of separate files for all the images that I've included in a particular book. By separating the files from the book, it allows me to manage the resolution of these files better and to catalog them more easily. Having separate files enables me to track all the photos that I use in a particular project. It also helps me to reuse some of these images if I need them in on a future project.

In summary, I have presented several tools that you can use to both write and format your books for both ebooks and print. I will talk a little bit more about the process of how to prepare your books for publishing in the next chapter. These are certainly not all of the tools that are available for publishing. As I come across more useful tools, I will add them to the list on my website as well as some blog posts outlining those that I think are useful. You can check these out on my website at:

https://destinlearning.com/toolbox

2.5 LAB Develop Your Book Outline

After you have created your book cover and you have your concept firmly in mind, it's now time to create your book outline. In this exercise, you will go through and create a high-level outline for your book so that you can use this as a baseline to start cranking out content one chapter at a time. This outline will set the tone for the remainder of the production process for you to create all of the remaining content for your first draft.

The first step is to develop a list of the major chapter headings. Work through the book from the beginning to the end. Match all of the goals you had in mind when you developed your book concept and use the cover to help flush out any ideas that you might have for this project. You can include any introductory material such as the preface, table of contents, and introduction. Try to be as complete as you can at this point. Also, be sure to include any post material for the book, such as information about the author, book summaries, and any supporting appendices.

Next, you will break down each of the chapters into any necessary subsections. I like to number these in a reporting style as in the following:

1.0 First Chapter
    1.1 First subtopic
    1.2 Second subtopic
    1.3 Last subtopic for Chapter 1

Even if you don't intend to have your table of contents read like this, I found it useful to have this type of numbering system. It helps me track my production process as I develop the material for each chapter and sub-topic.

Optionally you can take this to the next level and begin to flush out details for each subtopic. However, I generally leave this to the process when I start writing each chapter out, and I will flush out those details one subsection at a time as there as I write them. I like to jot down all of the ideas for each subsection any corresponding research on a piece of paper. I then use this as a basis for my dictation when I write my first draft of the topic. This process is, of course, up to you how you want to implement this. I have found that this process has increased my productivity. I recommend the most detail that you can put into the outline will help you later on when you're writing each chapter of the book.

I have included an example outline from one of the books that I am currently writing that you can use as an example. This book is a book on Microsoft Excel and how to use it for practical applications. I've included a small segment of this outline so you can get the idea in this book. But you can also download the Excel spreadsheet version of us if you find it helpful.

2.6 Help I'm Stuck!

"When I'm stuck in my writing, the world is amiss. If I'm eating a sandwich, it's an unsettled sandwich. If I'm in the shower, it's an incorrect shower. It's profoundly uncomfortable. But it's what keeps me pushing."
Melissa Rosenberg

In this business, one of the most dreaded problems is getting stuck on a piece of content. For some reason, particularly authoring books and written content, getting stuck can become a real crisis to get stuck on a chapter or areas of a book. It may be challenging to get started again. One of the most common problems is the blank page syndrome. It is often challenging to get started on a particular project. Many times once you get going, it is much easier to sustain progress once you get a decent start. In this chapter, I will share with you some of the strategies that I've used to help keep me on track and to produce content regularly. Some of these strategies may seem very simple, but they have been useful for me in creating a large amount of content and keeping on schedule.

One of the things I like to do is to start my book project by creating the book cover very early in the process. I also know a lot of authors who do this at the very end. For me, it helps to do this as soon in the process as possible so I can help visualize the entire project. I'm a very visual person, and using the image of the book cover helps me crystallize a lot of the ideas that I want to include in the book. This process helps me from getting stuck somewhere deep in the project.

Another tool that I use to keep me on track is to develop a schedule for my projects. It is much easier for me to chip away the project and do a little bit at a time and to sit down for massive marathon writing sessions. I try to schedule my projects so that I can do a little bit every day. By doing this, bit by bit, I can add one chapter here and there to the book so that I am making regular progress. I also found by doing this I'm now able to now produce content on those days where I'm not at my best. It is crucial if you are trying to develop a reliable income out of being an author to produce content regularly. Working on a schedule will help you develop the discipline to do this.

I also use post notes that remind me that today is a day that I scheduled for writing and jot down my assignment for the day. I know that this is really old school and low tech, but I like to have the reminder in my face to help motivate me.

Another trick I use is to include an image at the beginning of each chapter in my book. By doing this, I can help visualize the contents of the chapter and then begin flushing it out with a quick writing session. Many times I will flush out all of the images for an entire book before I go in and start writing out all the details of each chapter.

If you are stuck, you can try to change your venue. This change may help you transform your frame of mind. I like to visit the local library regularly to go and produce a particular piece of content. The local library in my area has a great shared working area and is very quiet but not so quiet that it is uncomfortable. There are also many people working in this area. Sometimes it's better to be around people that are working and doing things to help motivate you to get your tasks completed. Another benefit of this is I sometimes schedule my trips to the library to complete specific tasks. I don't leave until I have completed my goal for the day. This technique is a great way to keep you on schedule.

Another technique that you can use when you get stuck, is to begin dictating your ideas to your computer or even a tape recorder. I routinely use the voice typing function in Google Docs and use it to produce draft material that I will go back and edit later. For me, this takes the drudgery out of writing very long sections of content and allows me to be more creative. Since I can dictate to the computer, this frees me from the monotony of typing. This technique, for me, makes me extremely productive. The downside of this is your material will require substantial editing as the dictation process introduces a lot of subtle errors. I have found through experimentation that the voice typing function in Google docs is much better than the voice to text function on the Apple Macintosh, for example. It is also much better than some of the commercial software products that I've tried. The great advantage of the Google function is that it is free, and there is no charge to use it.

Another technique you can use is to find someone interested in the material you are producing and have a conversation about the item you are stuck on writing. One of the difficulties with this process is it may sometimes be challenging to find someone who wants to talk about the material that you're trying to develop. However, with some creativity, you should be able to find somebody to discuss the content. You can even find a blog somewhere where you could post something and get some ideas back from some people who like to follow that type of material.

Another variation of this technique is to sit down and interview yourself about a particular topic. Although you will feel exceptionally silly doing this, it may help you over the hump. If you have a nice private office like I do you can do these types of things without people laughing at you. I found the more creative you can get by introducing more processes into the equation will help you jump start your level of productivity. These techniques will allow you to produce high-quality content at a steady rate routinely.

Writing is a creative process, and it is difficult getting used to producing content regularly. When you first start writing, you will notice that you will have some sessions where you are incredibly productive and other times that you can't get anything done at all. With practice and dedication, you will be able to overcome this limitation and be able to produce content on a more regular basis and level out the production process with consistent, predictable writing sessions.

I hope that you'll never encounter a dry spell where you have difficulties producing content, but I'm pretty sure if you're like everyone else you will have your fair share of these types of problems. I hope that some of the tips in this chapter can help you out of some of those ruts, and then you will be able to maintain a high level of productivity. Good luck with your writing!

3.0 Publishing Your Book

"It took me fifteen years to discover I had no talent for writing, but I couldn't give it up because by that time I was too famous."
Robert Benchley

In this chapter, I will describe all the steps that are necessary to publish both your ebook and print book to Amazon Kindle. Although there are other platforms to publish your book Amazon is it is the largest self-publishing platform in the world. Before I describe those steps, I will describe some of the other platforms I have used to publish my books and some of the features of those platforms.

3.1 Where to Publish Your Book

There are many different places where you can self publish your book. In this chapter, I will describe some of the sites that I've used so far in my publishing business. Also, I will mention a couple of others that I've looked at but have not yet had the opportunity to use. I'm sure there are many other services as well that I haven't listed here. As I find more, I will post them on the resources section of my website so that you can get additional information as I move further in the process.

Amazon Kindle

Amazon is the world's leading seller of books. Amazon Kindle is Amazon's service for self-published authors. Amazon allows you to upload and sell ebooks as well as on-demand print books as well. Previously, Amazon had two different services for self-published authors that they have now consolidated under the Kindle umbrella.

Since Amazon is so massive, there are a lot of resources to support people that want to publish on the Kindle platform. There are many examples of courses that will teach you the in's and out's of authoring and selling books on Amazon. If you are going to make the effort to produce a video course, it is not much additional effort to format all that material as a book and sell the ebook online. I have personally found that I like to write a book first and then put produce a video course based on the material in the book. This allows me to deliver a much more complete course in a video format since I've already done all the research for the videos by writing the book.

I don't think this is a necessary step for all people. You may find it is perfectly adequate for you to produce course videos and not write an accompanying book. Many people genuinely dislike writing books and find that producing videos proves to be a much simpler process for them. I have discovered that I enjoy writing books and I also enjoy creating videos. I simply combine the two methods for my coursework.

The revenue share for Amazon for an ebook is either 30% or 70%. If you price your ebook under $9.99, you can get a 70% Revenue share from Amazon. The minimum price for an ebook on Amazon is $2.99. If you price your ebook outside of that range, you will get a 30% Revenue share. So for example, if you sell a book for $9.99 beneath the 70% Revenue share, you will get about $7. There are also distribution fees on top of the revenue split, and this is dependent on the size of your book.

They have an online calculator that will tell you your Revenue split before you publish your book.

Amazon pays its authors 60 days after the end of the sales month. Amazon automatically deposits the funds in my account that I set up during the registration process. You can track your sales on a sales dashboard daily can you can see the breakdown of sales by title, location, and print versus eBook. There is also a new feature on the website that will give you an estimate of the funds that you make for pages that are read from the KDP select program.

My experience with publishing on the Amazon platform has been very positive. The platform is straightforward to use, and Amazon gets an incredible amount of web traffic, so it is very possible to generate a large number of sales on Amazon. Through the use of Draft to Digital, I tried to go to a broader distribution earlier on in my publishing business. I had almost no success on any of the other platforms such as

Barnes and Noble and iBooks. Amazon commands a massive market share, and if you're starting all-out, it is probably the site where you will have the most success. I do think if you become a well-known author, it would be better to go to a broader distribution to more websites as you establish your name. The two channels that I concentrate on now are Amazon and Ingram Spark.

Draft to Digital

Draft2digital is a book aggregator. With a book aggregator, you can publish your book to one site, and they will, in turn, publish your book to a large number of online bookstores. Draft2digital currently publishes books to the following sites:

- Amazon
- Apple Books
- Barnes & Noble
- Kobo
- Playster
- Scribd
- Tolino
- 24Symbols
- OverDrive
- bibliotheca
- Baker & Taylor

Draft2digital will take a fee with each sale that you make and then send you a monthly check based upon your sales. If you publish to

Amazon directly, you cannot use Draft2Digital to publish the same book through this channel. Draft2Digital currently only works with ebooks and does not service hard copy books.

In addition to their book aggregation service, Draft2Digital also offers a simplified way to upload documents that you can quickly turn into ebooks. I have found, however, if you have a large number of Graphics or complicated formatting requirements, you will need to format your ePub book for distribution on this platform.

IngramSpark

IngramSpark is a book aggregator. If you wish to sell your books and get them into bookstores like Barnes & Noble, this may be the best approach to do that. With Ingram Spark, you can self publish a book, and they will list it in their catalog for availability to market to on-premise bookstores such as Barnes & Noble. IngramSpark can publish both ebooks and print books as well. With print books, they will print them on demand as their customers order them from their catalog. There is a fee of $49 for each book that you set up to be in the catalog.

IngramSpark will offer your books at discounted prices to attract bookstores to buy your books. There are printing and distribution fees for books and an online calculator to help you determine what your profit will be for an individual book. As a ballpark example, if you sell a book for around $15, you will probably make a profit of approximately $3 for each book that you sell.

The good news with this distribution and delivery is that you don't have to be involved in the ordering and shipping process. IngramSpark takes care of all that and will also help or your books to their potential buyers.

I have just recently started publishing via IngramSpark, and I'm happy with the results so far. It has created a second reliable income stream for my publishing business, and I plan to continue publishing with them well into the future.

Other services

There are also many other services on the market, such as iBooks, Smashwords, and Book Baby, to name a few. I have not included details with these services since I have not had a chance to use these platforms yet. As I find other services that have been successful for me, I will post updated information on my website.

3.2 LAB Sign Up for Your Kindle Account

In this section, I will show you how to sign up for your Amazon Kindle account. There are several screenshots, and I will show you all of the information you will need to provide to set up your KDP account. First, you'll need to navigate to:

https://kdp.amazon.com

This link is the homepage for Amazon Kindle. If you already have an Amazon account, you can sign in and then provide the rest of the account information that Kendall will need if you do not have an account click on the signup button that's displayed below.

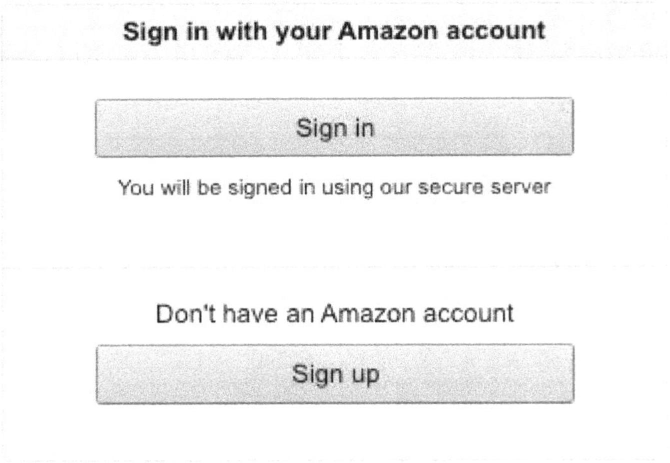

Sign in with your Amazon account

Sign in

You will be signed in using our secure server

Don't have an Amazon account

Sign up

Next, click on the Create your KDP Account button.

Sign in with your Amazon login

If you are new to KDP, you can use an Amazon login to register. Just sign in with your existing Amazon login or create a new account.

Email (phone for mobile accounts)

Password   Forgot your password?

Sign in

By continuing, you agree to Amazon's Conditions of Use and Privacy Notice.

☐ Keep me signed in. Details ▾

New to KDP?

Create your KDP account

Next, enter your name, email, and password for your account.

Create account

Your name

[   ]

Email

[   ]

Password

[At least 6 characters   ]

i Passwords must be at least 6 characters.

Re-enter password

[   ]

[Create your KDP account   ]

By creating an account, you agree to Amazon's Conditions of Use and Privacy Notice.

Already have an account? Sign in ›

Now that you have established your account, you will have to provide Amazon with your author/publisher information, your bank account information so you can get paid, and your tax information. The following screenshots are the details for those three pieces of information.

First, you need to add in your contact information, as in the following screenshot.

Author/Publisher Information
Please enter your address.

Country or Region	United States
Full Name (What's this?)	ERIC R FRICK
Address Line 1	12345 Test Drive
Address Line 2 Optional	
City	Someplace
State/Province/Region	Ohio
Postal Code	12345
Phone	999-999-9999

Next, you will need to add your bank account information, as illustrated in the following screenshot. In the first selection, you will need to select the location of your bank account. Following that, you will enter the account details, such as routing number, account number, and bank name. Once you have completed the form, click the add button.

The last step in the process is to enter your tax information. The easiest way to do this is to complete the tax interview that guides you through this step by step. You can start the interview process by clicking the button to launch the interview.

Tax Dashboard

Tax Information Interview

Your tax information has been validated. If you need to provide updated information, please click Take Interview.

Take Interview

After completing the tax information, your account is now complete, and you will now be ready to begin publishing!

3.3 Tips for Book Editing

"Editing should be, especially in the case of old writers, a counselling rather than a collaborating task. The tendency of the writer-editor to collaborate is natural, but he should say to himself, 'How can I help this writer to say it better in his own style?' and avoid 'How can I show him how I would write it, if it were my piece?'"
James Thurber

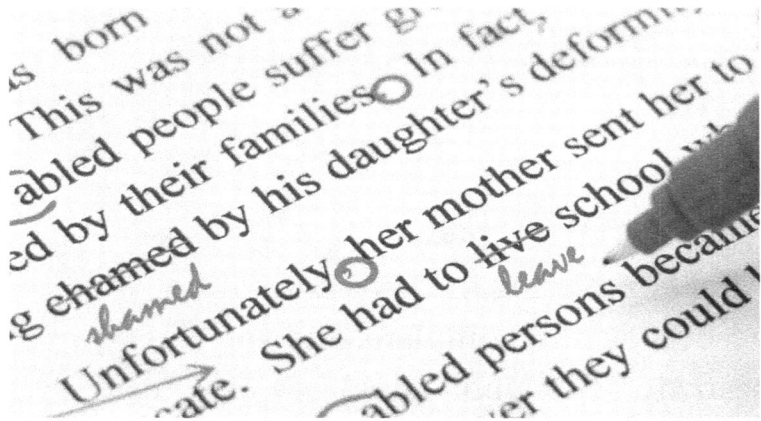

Quality editing is a vital step in producing a great product. If you publish a book that contains a large amount of grammatical errors and spelling mistakes, you'll indeed receive harsh reviews from your readers. However, editing is much more than merely correcting grammatical errors and spelling mistakes. Many different types of editors can help you with your book. The two that I suggest would be most useful are copy editors and developmental editors.

There is also a timing aspect with editing, and it depends on the type of book that you're publishing. The first aspect of editing I will discuss is that of a developmental editor. This type of editor will help with the structural elements of your book. Also, they can help with the overall theme and flow of ideas. If you are writing a fictional book, you will probably want this help earlier in the publishing process to make sure that your storyline flows well and the characters are well-developed. If you are writing a technical book, this is perhaps a beta test for your book. You can run this by other people that are experienced in that particular area to make sure that you're covering the topic thoroughly and representing the material in a way that is easy to understand.

I write mostly technical books, so the type of editing that I depend on the most is for peers to first review my writing. I can then make certain that I have not left anything out and that I have presented the material in a way that is easy to understand. I want to gain a sense from my peers that the book delivers on the concept that I provided to them earlier.

The other type of editing is more of a copy edit function. This type of editing makes certain that all of the sentence structures are correct, there are no misspelled words, and the punctuation is proper. I have found that if I use a platform such as Grammarly prior to delivering to the editor, I get a much cleaner end product. I use this tool every day, and it has helped me immensely. I highly recommend using the professional paid version of Grammarly to routinely run the text through the software before sending it to an editor.

The other aspect of editing is finding someone that you can work with comfortably. You want someone that will give you an honest opinion of your piece, but won't attempt to rewrite the book in their own style. You will have developed a style of your own, and you should stick with it because it's your honest voice. Sometimes you may find an editor that wants to really write the book in their own style, and that could become a conflict. It may take some trial and error to find the right editor for you.

3.4 How to Format Your Book

At this point in the process, you will need to format your book. You will need files to support e-books and print books. There are many options and different tools that you can use at this point, and I have included a section earlier in this book that describes these book writing tools. Rather than explain all the potential processes that you could use to format your book, I will describe the process that I currently use and why I use it.

I no longer use the same tool to develop manuscripts that I use for formatting for my final output. At one time I used Scrivener to do all of this, but I found myself spending too much time trying to manage the process and less time writing. Once I moved to Google Docs to manage my manuscripts, I found my productivity became much higher. I also had more freedom to spend more time writing and less time worrying about formatting.

Currently, I periodically export documents from Google Docs and download them to my Macintosh computer. I then import these documents into Apple Pages and format as an Epub for my ebooks. The next step entails making a copy of that document and setting the document size for my desired trim size for my print book. The final steps are paginating the document, and then exporting it as a PDF for upload to Amazon Kindle. This process may seem convoluted, but trust me, I have burned days upon days trying to find out the ultimate method to use and have tried a lot of tools. This process currently works best for me.

I realize that many people will not have a Mac and access to Apple Pages to use for this process. My next best recommendation is to download and install Kindle Create and then use the tool for final formatting for both ebooks and print books. The software is straightforward to use and integrates well with an Amazon Kindle.

My last recommendation if none of those methods work for you is to use Microsoft Word for everything. Word is perhaps the most widely used word processor in the world, and there are many examples on the Amazon platform that can show you the houses of how to format books for input as books. However, I had a lot of difficulty with this process and moved on after several tries. Some of their import capabilities may be better now based on word templates. I'm delighted with the process that I'm currently using.

One final option is to hire out the formatting process and use a service to do this. There are many reasonably priced options on services like Fiverr, where they can do this for you. I operate my business on a fairly tight budget and have decided only to hire out the book cover process and do the remainder of the work myself. This process may change sometime in the future if my sales increase, but for the current period of time, I do the formatting myself. You may find that this is a process that you don't enjoy, and that you will be better off having someone else do it for you.

3.5 How to Self Publish Your eBook with KDP

"I had a period where I thought I might not be good enough to publish."
Stephen King

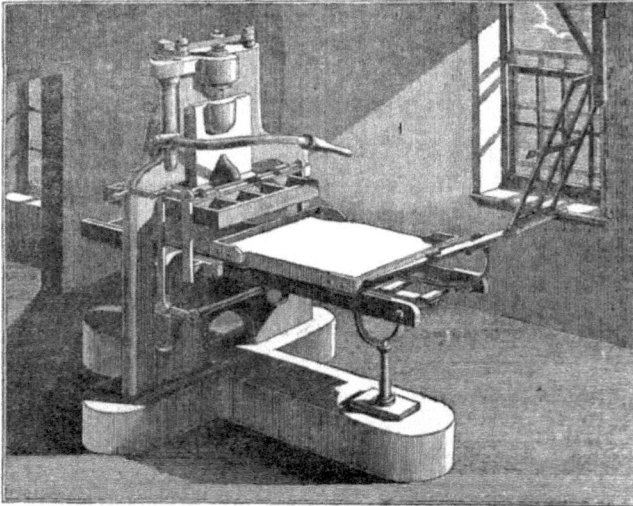

In this section, I will give you the step-by-step instructions on how to publish your ebook on Amazon Kindle, or KDP. I will describe all of the fields that you will need to complete the forms for the information that you need to upload to publish your book. So let's go ahead and get started.

The first step in the process is to log into the KDP account that you established earlier in this chapter. Once you navigate to: https://kdp. amazon.com , click on the sign-in button and then enter in your username and password.

Self-publish eBooks and paperbacks for free with Kindle Direct Publishing, and reach millions of readers on Amazon.

Once you are logged in it will take you to the Bookshelf page. At the top of the bookshelf page, you can then click the button to add a new Kindle ebook as in the following figure.

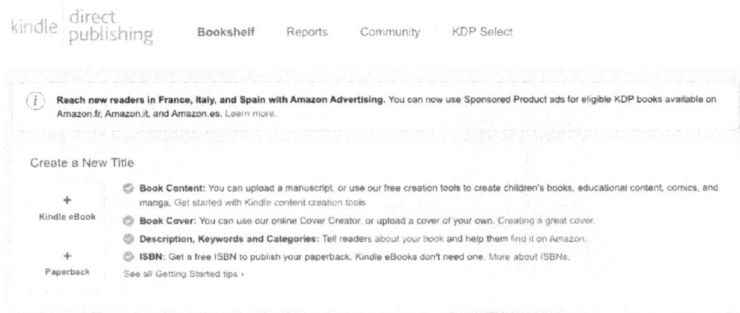

The book information is broken down into three major tabs. These are:

- Book details
- Book content
- Book pricing

The following screenshot displays these tabs. You will need to complete all of these before you can publish your ebook.

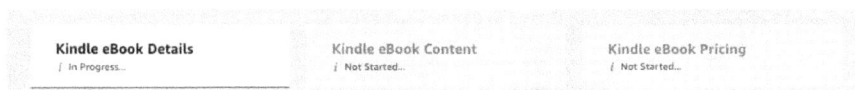

Kindle eBook Details
/ In Progress...

Kindle eBook Content
/ Not Started...

Kindle eBook Pricing
/ Not Started...

In the following pages, I will describe in detail each of the fields within each of these three tabs.

Let's first look at the Kindle eBook details tab.

The first part of the Kindle ebook details is very straightforward. The first thing to do is to select your desired language. This entry indicates the language that you used to write the book. Next, you will enter the book title and subtitle. Note that the subtitle is optional. Following that, you will enter the series information. If this book is part of a series, then enter the name, if it is not, you can leave it blank. The same goes for the edition if this is a first edition you can also leave this blank. The last information to enter is the author's first name and last name.

Next, I will describe the middle portion of the book details form that is displayed below.

Contributors	**Contributors** (Optional)			
	Author ⇅	First name	Last name	Remove
	Add Another			

Description	This will appear on your book's Amazon detail page. Why do book descriptions matter? ▾
	Fill this in later. Fill this in later. Fill this in later. Fill this in later. Fill this in later. Fill this in later. Fill this in later. Fill this in later.

3821
characters left

Publishing Rights	◉ I own the copyright and I hold the necessary publishing rights. What are publishing rights? ▾
	◯ This is a public domain work. What is a public domain work? ▾

Keywords — Enter up to 7 search keywords that describe your book. How do I choose keywords? ▾

Your Keywords (Optional)

self publishing	self published authors
self publishing on amazon 2019	how to become an author
work from home	self employment books
write your first book	

If there are additional contributors to your book, you can enter their name in this section. Following that, you will enter a detailed description of your book. This description is critical, and Amazon will use this as the main topic on the sales page for your book.

Following that is a checkbox that you can indicate that you own the copyright and hold the necessary publishing rights. There is also an additional selection that this is a public domain work. If you have further questions on these fields, there are links that you can click to get additional details.

The last section in this group is the keywords that are associated with your book. These keywords have a significant effect on how the website finds your book when people are searching amazon.com for books. There are many different techniques that you can use to generate good keywords so that your book will get traffic. I've included some links on some excellent articles that I have found and used to help develop my keywords. This process can be a challenging and frustrating subject for new authors. The good news is since there is no charge for revisions, you can change these keywords over time if you find your initial sets not working very well.

The following screenshot is the last data that you need to enter in the book details tab.

The first selection is to associate your book with up to two different categories. You can look through the list and determine which categories are best suited for your book. To select this, you will probably need to do some searching through Amazon for different books that are similar to yours and see what categories are listed. That's perhaps the best way that I found to set up these categories.

The next selection is to enter a minimum and maximum age group if that applies to your book. If there is no real age range for your book, you can leave the default selections and move on to the pre-order choice.

The last selection is whether you intend to release this book is available for pre-order or if you're ready to publish your book. In most cases, you will select that you are ready to release your book. You would use the pre-order if you're an established author, and you would like your readership to have advance notice that you are about to publish a book.

After you have completed all of this information, click the Save and Continue button, and the system will move you to the next tab, which is the content information.

The next section contains the book contents. The following screenshot is the top portion of this page.

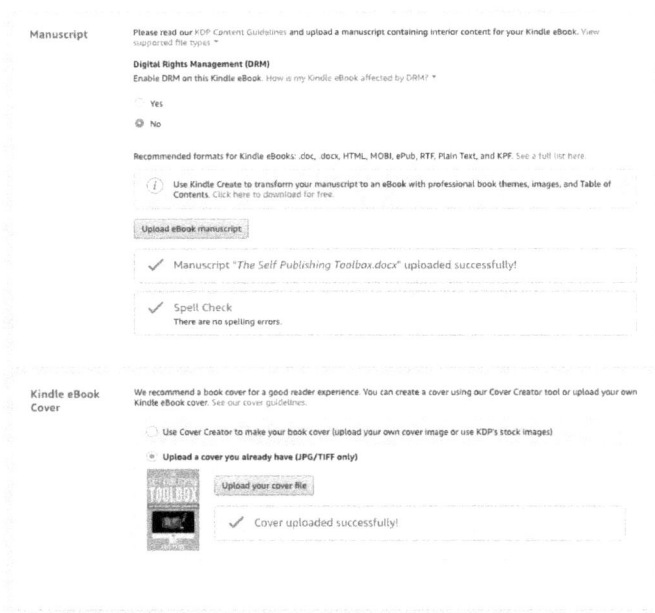

In the first subsection of this page, you will be able to upload your ebook manuscript to Amazon. You will note on this screen; there is a list of recommended formats better supported. Probably the most popular format is a .docx file that is a Microsoft Word file. I found it difficult to control the formatting with a Word file and so I have moved to use epub files, which I find convert much better. You can also upload a kpf file, which is the output from the Kindle Create program. This program is also an excellent option, especially if this is your first book. The Kindle Create program supports formats for both ebooks and print books.

In this subsection of this page is an area to upload your ebook cover. There is a link that has more information about the specifications for the ebook cover format. Also, you can select to use the built-in Cover Creator to build your covers online and use some of Amazons' predefined cover templates. My advice is for you to try this to rough out some ideas. Once you find something close to what you want, then make a screenshot of this to use as a template to custom design your cover offline and upload it later.

The following screenshot is of the last two sections on the book contents tab.

The first section presented above contains a button where you can launch the online previewer. To do this, click on the launch previewer button, and the e-book preview or screen will come up. Using this, you can check how your final product will look to your readers as an ebook. You will want to check to see that the pagination is correct and that all the images are clear and formatted correctly. You will also want to verify that the table of contents has been uploaded and successfully constructed by Amazon. You will probably need to do this several times to fix any mistakes before you finally publish your book.

The last section presented above is for you to enter an ISBN if you choose to do so. If you are publishing on multiple platforms, this is a necessary step; if you are only publishing on Amazon, Amazon does not require an ISBN for ebooks. There is also an optional entry for a publisher as well. Click the Save and Continue button to move on to pricing. The screenshot below is the first half of the pricing section.

KDP Select Enrollment	**Maximize My Royalties with KDP Select** (Optional) With KDP Select, you can reach more readers, earn more money, and maximize your sales potential. Learn more about KDP Select. How Do I Enroll? * ☑ Enroll my book in KDP Select
Territories	Select the territories for which you hold distribution rights. To enter the Kindle Storyteller contest, you need make your book available at least in Amazon.co.uk. Learn more about distribution rights. ⦿ **All territories (worldwide rights)** What are worldwide rights? * ◯ Individual territories What are Individual Territory rights? *
Royalty and Pricing	**KDP Pricing Support (Beta)** See the relationship between price and past sales and author earnings for KDP books like yours. [View Service] **Select a royalty plan and set your Kindle eBook list prices below** ◯ 35% ⦿ 70% ⓘ Your book file size after conversion is 2.18 MB.

Primary Marketplace	List Price		Rate	Delivery	Royalty
Amazon.com ⇕	$ 3.99	USD	35% *	$0.00	$1.40
	Must be $2.99-$9.95 * All marketplaces are based on this price		70%	$0.33	$2.56
Other Marketplaces (12)					⌄

The first selection you will make is the checkbox, whether or not to enroll your book in the KDP select program. If you select to participate in this program, you enter into an exclusive relationship with Amazon for your book. In return, you get paid for people reading your book that are members of the Amazon KDP program, and you also get access to a higher pricing level where you get a 70% revenue share. The last entry on this page is the pricing. For you to participate in the

KDP select program, your pricing must be between $2.99 and $9.99. You can go outside this range, but then you can no longer participate in the Kindle Select Program.

The bottom half of this form is the very last portion of this process. The book lending program and the terms and conditions fields are displayed below.

In this last step, you will select whether you want to participate in the book lending program. This program allows customers to lend your Kindle book after purchasing to their friends and family for 14 days. I generally accept this as one of the conditions for publishing. The last bullet outlines that you agree to Amazon's terms and conditions to publish your book. There's nothing really to select in this box. Once you've completed all of this information, you can save it as a draft if you still want to make more changes. The other option at this point you can click on the Publish Kindle ebook button and the system will submit the book to Amazon for review and then subsequent publishing. If they don't find any issues, they will approve it and notify you. I found this review process is relatively quick and generally happens within 24 hours.

3.6 LAB Publish Your eBook with KDP

In this exercise, you will go through the steps outlined in the previous section and set up your ebook in KDP for publishing. Use the account that you set up earlier in this book for your publishing.

Even if you are not complete with the final version of your first draft yet, go ahead and go through this exercise and set everything up. You can use whatever you have to date for a draft as a manuscript, but go through all the details and set up everything else as though you're going to publish your book.

Now is an excellent time to check your formatting to make sure it is uploading to Amazon properly. Use the book preview feature on the KDP upload page so that you can see how your book is going to look when people buy your ebook. I found that there are always surprises and how the book looks after it is imported, and it's a good time for you to experiment with a little bit of trial and error to make sure that you're getting the results that you want. I have had the best results uploading epub files, and they seem to upload very clean. You can also upload other formats as well, but I have had the best results with epub files for the types of books that I write.

If your manuscript is not ready to go at this point, save this as a draft and then upload subsequent versions of your manuscript as you make changes. By setting this up now, once you're ready to publish, you will only have to upload an updated manuscript and then hit the publish button. It generally takes about 24 hours for Amazon to review your book before it goes live.

This process is also how you will upload updates whenever you want to make changes to your ebook. There are no charges for updates, and you only need to upload an updated manuscript and then hit the publish button one more time. Updates also go through a speedy review process from Amazon.

3.7 How to Publish Your Print Book with KDP

The process to publish or print book with Amazon Kindle it's almost identical to the e-book publishing process. If you've gone to the trouble to write and publish an ebook, it's really a no-brainer to publish a print book as well. The publishing process goes through almost the exact same user interface as the e-book process.

The significant difference with the print book process is you will have to select a trim size for your book. Once you have done this, you will have to format your book as a PDF file that matches that trim size. Once you've completed that process, you simply upload your PDF file to the Amazon website and preview to make sure all the formatting is correct. The other difference with uploading a print book is the preview function is different and shows you how a print book would look in person. It is a straightforward process, and once you have uploaded your ebook, you will have no trouble uploading a print book.

In my experience, it has been well worth my while to publish print books. My print books account for 60% to 70% of my sales every month. You would think in today's age of technology that no one is interested in print books, but that does not seem to be the case.

3.8 LAB Publish Your Print Book with KDP

This exercise will be very similar to the process that you used for the ebook setup. You will use the same keywords and descriptions as used for your ebook to publish your print book. The setup pages are identical to the ebook process to publish your print book on Kindle with one major exception. The major exception is you will set some options for your print book such as:

- The color of the pages
- The trim size
- Options for the cover of the book

You will need to format both the cover in the manuscript for the desired trim size. I usually use six by nine inches for most of my print books. However, I have used other formats on occasion. Select the best size that makes sense for your particular book. If you are producing a textbook, you will probably want a larger size book for readability. If you're creating a novel or a fictional work, you may want a smaller trim size.

You will be able to download a template from Amazon once you select the trim size, and you need to input the number of pages as well for your book. Amazon uses this to produce the template. They will give you the amount of space that is available on the spine to put your title of the book on. The template is a PDF file, and you can use this in a tool like Photoshop to then overlay your book cover and make sure that it is appropriately formatted. If you do not want to do this or don't have the capability, you can use the same resource that I have recommended earlier in this book on Fiverr, and they can do this for you.

The other thing that you will need to do is to format your manuscript for the desired trim size. I do this in Apple Pages and set the page size to be the same as the trim size and put in the necessary page breaks based on the new trim size. Once I have completed this, I will upload a PDF file to Amazon. I also review the formatting using the print preview section of KDP to make sure everything is correct.

You will use the same process to publish revisions to your print book as well. There are no charges for revisions for your print book. Amazon employs a print on demand system once your book sells. By doing this, nobody has to have an inventory of books on hand. New books and revisions have typically taken in 24 hours for Amazon to review before they're ready for sale. Once you have this setup, it's a simple process to publish revisions.

4.0 Marketing Your Book

4.1 Where to Market Your Book

"Don't blame the marketing department. The buck stops with the chief executive."
John D. Rockefeller

I think the quote is fitting for this chapter that puts marketing your book into context. Although marketing is a critical step towards success, you can't market a product that is fundamentally flawed in some way. If you focus on producing your best work and have done your due diligence with the editing process, you can then move on to marketing your book.

Fortunately, there are many tools available for the self-published author that make this process much easier than it used to be. Several paid services can assist with book marketing, as well. However, I have not used any at this point and can't comment on their effectiveness. At this point, my publishing business is a modestly sized endeavor, and I have decided to reinvest most of the income I currently make with this into producing more content rather than professional services.

If you decide to publish your book on Amazon Kindle, one of the first decisions you will have to make is whether or not to sign up for the KDP select program. When you sign up for this program, you will enter into a contract with Amazon to exclusively publish your book on Amazon and not on other publishing sites. I should note that this program is for ebooks only and not for print books.

In return for your contracted exclusivity, you gain several benefits by being in the KDP select program. The first benefit is,you will be able to run periodic free book giveaways to help stimulate sales of your books. With the rules of this program, you can do this every 90 days. Also, you will be able to have periodic sales called a Kindle Countdown in which you can discount your books for a 7-Day. With a Kindle Countdown, you can start with a more substantial discount and then step up for smaller discounts in one or two-day increments. I have used both of these programs, and they've been successful in helping me generate sales for my books.

The other benefit you gain with the KDP select program is that you get paid for the number of pages read on your ebooks through readers that loan the books that are part of the paid program that gets free access to books on Kindle through a monthly subscription. Amazon pays authors based on the number of pages that are read by your readers and the amount that is in the Kindle select pool for that month. If you get quite a few pages read by your readers every month, that can amount to a nice monthly income. For me, if you are not a well-known author, this program probably offsets any additional sales you would get for marketing to other websites such as iBooks, Barnes & Noble, and other sites. This program renews every 90 days, so you could try it out for a little while and see if you like it. If you don't, then you can withdraw from the program and distribute your book to other channels later.

Another benefit of listing your books on Amazon is amazon.com has a built-in advertising capability for your books. You can create advertising campaigns directly from the KDP site that will market your books on the Amazon website to readers that are looking to buy. This type of advertising has several benefits. The first benefit is that readers are already looking to purchase books and are already on the Amazon website when they see your ad. The other colossal benefit is Amazon has a built-in One-Click buying mechanism that makes it easy for potential readers to buy directly from your ad.

You can track the performance of your advertisements with a metric called ACOS, which stands for Advertising Cost of Sales (ACoS). You can look at this as a percentage and judge the effectiveness of your ads. For a simple example, let's say you sell $100 worth of books from your ads, and it cost you $50 to run those ads you would have an ACoS of 0.5. That would mean for every $50 you spend, you would generate $100 in revenue. The goal is to have the smallest ACoS number possible. Anything ACoS over 1.0, you are losing money at that point. In the next section, I will show you step-by-step how to create an Amazon campaign for your book.

Another site that you can quickly generate ads for your book is Facebook. Facebook allows you to have highly targeted and customizable audiences for you to display your ads. The Facebook advertising system is integrated directly into the Facebook website. It is straightforward to set up. You can also have a post that you can turn instantly into an ad by boosting that post. You will need to have a business page to take advantage of Facebook ads, which is free to set up. Although the Facebook ads are straightforward to set up and run, it takes some practice; it is challenging to get Facebook ads to pay off effectively. To date, my Amazon ads have been much more profitable than Facebook ads for selling my books.

YouTube is another site where you can run ads for your books. I highly recommend that you start a YouTube channel if you intend to make this a long-term business. Google owns YouTube, and it is the second-largest search engine in the world. You have the chance to get a lot of exposure by creating videos about your books. It's also a chance for you to provide additional material for your books publishing YouTube videos for your book subscribers. In the trailer of your YouTube video, you can always put a brief advertisement about your book. You can also set a link in the video description of how to purchase your book on Amazon or other websites.

There are many other websites for marketing your books, such as paid services, Google Ads, Bing Ads from Microsoft, and many others. Many paid services will help you market your book for a fee. I have not used any of these today, but if I have relied mostly on YouTube and Amazon advertising help market my books. I highly recommend you start with

Amazon advertising first since it is the easiest and most effective and then branch out into other marketing strategies following that.

4.2 LAB Create an Amazon Add For Your Book

In the section, I will show you the step-by-step instructions on how to create an ad for your Amazon book within the Amazon Kindle. I will base this example on a book that I have already created and show you some of the options that you can select to get started. I will assume in this example that you have already signed up for a KDP account. Also that you already have a book on your bookshelf that you have published. I will use one of my existing books as an example.

First, click the Promote and Advertise button on the desired book on your bookshelf.

Next, select the desired market place and then click the Create and Ad Campaign button.

Run an Ad Campaign

With Amazon Advertising, you set your budget, targeting, and timing. You pay only when shoppers click your ads. To create an ad campaign, choose the Amazon marketplace where you want the ad to appear. To advertise this book in multiple marketplaces, repeat this step for each marketplace. Learn more ⌄

Choose a marketplace:

Amazon.com ⌄

Create an ad campaign

Next, select the desired campaign type. For this example, we are going to create a sponsored product ad. These ads are much simpler to create, and you can start with a lower investment. You can run these ads for as little as $1 per day. The minimum for a lock screen ad is $100; it is a bit more complex to set up. I recommend starting with the sponsor product ad and then expanding from there.

Choose your campaign type

Sponsored Products

Promote products to shoppers actively searching with related keywords or viewing similar products on Amazon.

Continue

Lockscreen Ads

These ads are based on shoppers' interests and are shown when they 'unlock' their Kindle E-readers or Fire Tablets to begin reading or shopping for books.

Continue

Now you will fill out some necessary information about your campaign. First, you will enter in a name for your campaign that will be easy for you to track once you go back to the system. Following that, you can select a portfolio for your campaign. You can group many campaigns into a collection to easier track them. Since this is your first ad, you will not have any portfolio to select from, so you can leave the default settings. Next, you will enter his start date and end date. You don't want to have a particular end of the campaign you can use the default settings. Next, you will enter a daily budget for your ad. I will note that this daily budget represents a ceiling of how much the maximum daily spend will be. It has been my experience, however, that you rarely reach the limit. For example, if you have a daily budget of $5, you will be lucky to spend one or two dollars every day on the ad. The next setting is based on targeting keywords. You have two choices of automatic targeting and manual targeting. With manual targeting, you will enter your own keywords, and this probably represents the best way to structure your ads. However, to get started, I recommend you use automatic targeting just to get started and get some experience.

Create campaign

Settings

Campaign name

Example Campaign for Book

Portfolio

No Portfolio

Create portfolios to organize campaigns, set budget caps, and track performance.

Start   **End**

Jan 3, 2020   No end date

Choosing no end date means your campaign will run longer, and a longer timeframe can give you better insights on search terms and keyword performance to further optimize your campaign.

Daily budget

$ 5.00

Most campaigns with a budget over $5.00 run throughout the day.

Targeting

⊕ Automatic targeting
Amazon will target keywords and products that are similar to the product in your ad. Learn more

Use this strategy when you are first getting started or want to launch a campaign quickly.

Manual targeting
Choose keywords or products to target shopper searches and set custom bids. Learn more

Use this strategy when you know which keywords deliver the most value for your business.

For the next two settings, select the default settings — select Dynamic Bid for the Bid Strategy and select Custom text ad for the Ad Format.

Campaign bidding strategy

○ Dynamic bids - down only
We'll lower your bids in real time when your ad may be less likely to convert to a sale. Any campaign created before April 22, 2019 used this setting. Learn more

Dynamic bids - up and down
We'll raise your bids (by a maximum of 100%) in real time when your ad may be more likely to convert to a sale, and lower your bids when less likely to convert to a sale. Learn more

Fixed bids
We'll use your exact bid and any manual adjustments you set, and won't change your bids based on likelihood of a sale. Learn more

˅ Adjust bids by placement (replaces Bid+)

Ad Format

○ Custom text ad
Add custom text to your ad to give customers a glimpse of the book. Limit one product per campaign.

Standard ad
Choose this option to advertise your products without custom text.

In the last step, you will enter the custom text for your ad as in the following figure. After you have completed this, click the Launch Campaign button.

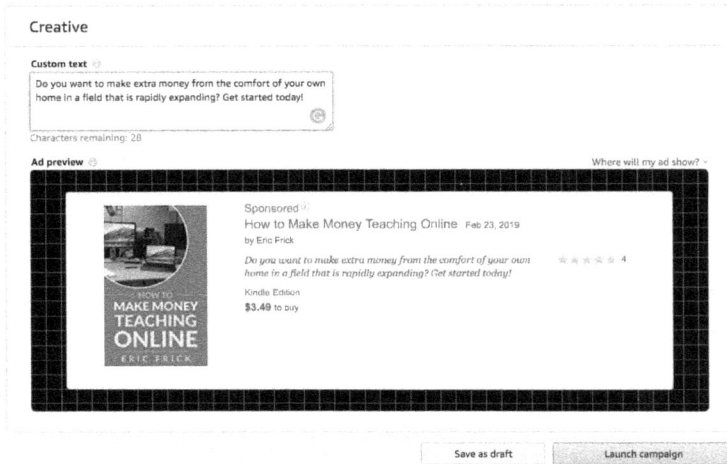

Once you have submitted your campaign, it will go through a review process by Amazon. Amazon will notify you once they have approved the ad. The typical approval process is usually a few hours. You can now track the ASoS on the add summary page to see you the ad is performing.

5.0 Summary

5.1 More Products You Can Offer

"I think we're having fun. I think our customers really like our products. And we're always trying to do better."
Steve Jobs

Once you have written your book, you can publish your material on multiple channels to help diversify your income. Your book represents a complete work on the subject and is an excellent baseline for other forms of the product to help you gain traction on multiple channels. In the next few sections, I will outline some of the possible different ways that you can publish your material on and some of the specifics on each one of those platforms.

The first platform I will talk about is having your book produced as an audiobook. Audiobooks are gaining in popularity and represent a great marketplace for your book. You will need to determine if your book is the type of book that is amenable to being presented as an audiobook. Not all books are a good fit for audiobooks. Some examples of these might be textbooks, cookbooks, or other types of how-to books, but many books are very easily translatable to the audiobook format. Another advantage of audiobooks is there is not as much competition on the audiobook side as with print and e-books.

Currently, there is a much higher demand for audiobooks than there is for supply. If you use a platform such as Amazon ACX, it is straightforward to convert your book to an audiobook. You have the choice of either producing the book yourself or hiring it out to someone to read the chapters of the book. The option is really up to you. ACX has a site that makes it very easy to hire someone to produce your book. The rates are relatively reasonable for someone to do it. If you are comfortable recording your own audio and it's certainly something you can do reasonably quickly once you've produced your book.

Another way to expand your revenue from your book is to turn your book into a video course. Since you've already done all the research for your material and have written chapters, it's pretty straightforward to produce matching PowerPoint slides and then record those as videos to present your content. Again not all books will be suited to be a video course. Still, with a little imagination, many publications will make an excellent video course. There is currently a robust market for video courses and many different sites where you can publish them. Two sites that I have used are Udemy and Skillshare.

One of the major sites for video courses that you can sell, produce, and publish free of charge is Udemy. Udemy has courses of all types, such as technical courses, IT courses on how to program, program management, cooking, and so on. All of these classes are reasonably priced, and Udemy claims to have millions of subscribers. Udemy has a revenue-sharing model that allows you to receive revenue for courses that are sold by Udemy.

Udemy also has a portion of the model if you market your own class as you get a higher percentage of the revenue. Even though course authors list courses at many different price points, most courses sell for $9.99, and you will get 50% of that revenue. So for planning purposes, you can count on about $5 of income for each course that you sell. Although this is not a particularly high price point, if you are selling a good number of these courses each month is an excellent way to supplement your income. Some course authors on Udemy have fantastically large followings and sell thousands of courses per month. You will probably not achieve that type of success early on, but it is an excellent way to add to your book revenue.

Skillshare is another popular video course site. Skillshare is a membership site that, as an author, pays you for every minute that students view your course. I have found that this marketplace is not quite as large as you Udemy but still offers an excellent chance for you to earn recurring revenue based on the work you completed with your book. Also, if you produce a course for Udemy is very little work to upload the videos to other sites such as SkillShare.

There are also many other sites that you can publish video courses on besides Udemy and Skillshare. Many more sites are being added by new companies every day. At the time of this writing, I've only published courses to Udemy and Skillshare and have not yet published to any other platforms. If I get more information on these sites, I will post that information in the appendix of this book and also on my website where you can get this additional information.

The last area I will talk about to publish your book material is YouTube. Google owns YouTube and is the second-largest search engine in the world. If your book has any type the value in being presented as videos on YouTube, I highly recommend that you start a YouTube channel.

Creating a YouTube channel has been one of the best decisions I have made to publish my content. It has taken me several years for me to grow a YouTube channel to reach a decent size. Now that I have taken the time to do this, I'm happy that I did, and it represents my best marketing channel to present my material. It also earns a small monthly revenue that helps add to my bottom line. Currently, this is not a large amount of revenue, but it does recur every month with very little work. This revenue represents an excellent passive income opportunity.

In summary, I have presented some of the other channels that I have used to publish my book works. I have been successful in using these channels to help grow and diversify my business. One of the benefits I have found this is if sales slump in one particular channel during the month, sometimes one of the other channels will earn enough money and compensate for this. This diversity levels out my monthly revenue for my business and allows for a much more predictable revenue stream. I continue to look for other channels to publish my work, but I use my books as the basis of truth for all of the different forms of my content.

5.2 Change the World!

"Education is the most powerful weapon which you can use to change the world."
Nelson Mandela

This is the third time I have included this chapter in a book. I have not cut and pasted this chapter from previous books (ok the image is), but I have taken the time to think this through and recreate it each time. This chapter is easily my most favorite material to write to describe the best part of being a content author. Along with the many advantages and flexibility of being a self-employed self-published author, some intangible benefits make being an author a very satisfying and fulfilling career.

In the opening section of this book, I suggested that if you are becoming an author purely to make money, you will be profoundly disappointed. Although making money is great and is necessary in today's world to help to pay the bills, there is indeed a much greater joy in your chosen occupation. This joy comes through helping other people. All of this may sound a bit trite, but this has been the motivation that has kept me going through difficult times. There were times when I thought I might throw in the towel and give up publishing forever. When you first start publishing, it will be a challenging experience. You will get some very critical and harsh reviews. At some point, however, this will be offset by some of the few people that you genuinely touch. Helping someone in their life is a tremendously satisfying job benefit.

I work as a content author for Linux Academy making video-based IT certification courses. As a part of this job, I regularly attend IT trade shows where we have a booth to announce new courses and training services. At almost every event, we have students who will come up to the booth and explain how our training has positively changed their lives. They will talk about how getting an essential certification led to a promotion at their current company of how they were able to launch a new career after taking some of our courses. It is an unbelievable feeling to have made a positive impact on someone's life, and to be able to help them through some of the content that you produced is incredibly gratifying.

One of Linux Academy's core values is "We are committed to changing the world by changing lives." To work at a company that truly puts their students first is a great environment to work in every day. It is a refreshing change from many of the companies where I have worked in the past whose only focus was on quarterly earnings and stock prices. I'm not saying those things are necessarily bad; we all have to eat and make a living. But sometimes your work can also contribute to a higher calling and meaning.

While I don't think that with one book you will dramatically change the world, you never know and it might! . I do believe that publishing will allow you to find your voice and ignite a passion that you can then tell the world with your message and in your own unique style. With platforms like Amazon and YouTube, you can deliver your content to a worldwide audience from the comfort of your own home, or wherever you want. The potential reach of your message is unlimited.

Likely the biggest hurdle you will have to face in publishing content is yourself. We all have a fear of failure and want to avoid any ridicule that we may face. When you write and publish a book for the first time, it places that fear at the forefront. You will find all kinds of excuses not to push the publish button for the first time. Even if you have completed the book and gone through the editing process, it is common for us to want to perfect the work before it goes out.

At that point, you have to find a way to "power through" and push ahead. We all have something to say, why not begin your journey now? Why not you? We all have our unique message, and books and online content are the perfect vehicles to deliver your message and story to the world.

In those times of self-doubt, I put the blinders on and forge ahead. Many times I find myself overplanning and not working on content. While planning is good, ***it can't be a substitute for pushing out more content.*** When you feel yourself becoming overwhelmed by all of this, go to the keyboard and write! You will then remember why you wanted to do this in the first place.

I have found that I have a few venues that I can count on where I can consistently produce content regularly. I am putting the finishing touches on this book while staying at my mother's beach place in Destin, Florida. Thanks, Mom!

I love writing while I am in Destin, the sound of the ocean is very calming, and I can really get deep into the material. You may not have access to a beach place or access to some exotic location. Still, I'm sure with a bit of imagination—you can find a place that inspires you and can help you during times where you have some difficulty getting material out the door. To help inspire you, the photo below is off the deck of my Mom's place in Destin, Florida.

I hope that some of these words will help inspire you during those difficult times of being an author. Writing and publishing content on your own is a much different type of job than a traditional 9-5 job and requires some unique motivation. You will need to be able to crank new material on those days where you would rather not. With some work and practice, you can develop a system that will allow you to do this. Experiment with small setup issues that can work to help you along the way.

You can do this! Start with writing your first paragraph, then the first chapter, and keep going. Don't let anything stop you. Take your first steps and change the world!

5.3 About the Author

Eric Frick

I have worked in software development and IT operations for 30 years. I have worked as a Software Developer, Software Development Manager, Software Architect, and as an Operations Manager. Also, for the last five years, have I taught evening classes on various IT related subjects at several universities in the central Ohio area. In 2015, I started Destin Learning http://destinlearning.com, where I have published several books and videos on technology and business-related topics. Please visit my website to see the complete listing of all of my publications.

I also work as a Cloud Training Architect for Linux Academy (https://linuxacademy.com), developing cloud-based certification courses. If you are interested in Cloud or Linux related certifications, please visit the Linux Academy website for more information.

5.4 More From Destin Learning

Destin Learning

Thank you so much for your interest in this book. I hope it has given you a good start on your path to becoming a successful self-published author. You can see more from my YouTube channel, where I am continuing to post free videos about software development, publishing, and entrepreneurship. If you subscribe to my channel you will get updates as I post new material weekly:

https://youtube.com/destinlearning

You can also sign up for my newsletter at https://destinlearning.com, where I will send out updates on new material. Thank you again, and good luck with your future with Information Technology!

Appendix Resources for Authors

Book Writing Tools

Scrivner
https://www.literatureandlatte.com/

Kindle Create
https://www.amazon.com/Kindle-Create/b?ie=UTF8&node=18292298011

Grammarly

https://www.grammarly.com

Google G Suite

https://gsuite.google.com

Book Publishing Sites

Amazon Kindle
https://kdp.amazon.com/

Draft to Digital
https://draft2digital.com/

Book Baby
https://www.bookbaby.com/

Ingram Spark
https://www.ingramspark.com/

Smash Words
https://www.smashwords.com

Stock Images

Shutterstock - This is the main site I use
https://www.shutterstock.com/

iStock - Alos really good paid site
https://www.istockphoto.com/

StockSpan.io - I have used some really good
free images from this site.
https://stocksnap.io/

Book Services

Pro eBook Covers
https://www.fiverr.com/pro_ebookcovers?source=order_page_summary_seller_link